My Wonderful Adventures
with
JESUS

Tom Sawyer

ISBN 978-1-64258-019-8 (paperback)
ISBN 978-1-64258-020-4 (digital)

Copyright © 2018 by Tom Sawyer

All rights reserved. No part of this publication may be reproduced, distributed, or transmitted in any form or by any means, including photocopying, recording, or other electronic or mechanical methods without the prior written permission of the publisher. For permission requests, solicit the publisher via the address below.

Christian Faith Publishing, Inc.
832 Park Avenue
Meadville, PA 16335
www.christianfaithpublishing.com

All scripture quotations, unless otherwise indicated, are taken from the *New King James Version*.
© 1982 by Thomas Nelson, Inc. Used by permission.
All rights reserved.

Scripture taken from the Holy Bible, New International Version®. NIV®
©1973, 1978, 1984 by International Bible Society.
Used with permission.
All rights reserved.

Cover design by Charlotte Rahrig
Cover illustration is protected by the 1976 United States Copyright Act.
© 2000

Printed in the United States of America

CONTENTS

Introduction..5

Chapter 1: Life That Truly Is Life..................................9
Chapter 2: Called and Equipped...................................33
Chapter 3: Important New Friends................................55
Chapter 4: Living in a Hostile Environment..................76
Chapter 5: The Power of Tracts...................................106
Chapter 6: A New Direction..141
Chapter 7: Time to Return..166
Chapter 8: Witnessing in Public Schools.....................185
Chapter 9: God's Unlimited Resources........................205

Conclusion...221
Witnessing Tract..224

INTRODUCTION

I grew up in a small community in the mountains of western Pennsylvania. Our village was too small for a church, but a Presbyterian Church in a nearby town started a mission there, which I attended throughout my childhood.

One Saturday afternoon as I stood looking out the window of the mission, a strong impression came to me: "Fisher of men." At the time, I didn't know what it meant, but now, I know that it was the Lord Jesus telling me what He had called me to do.

Some years later, I sensed two desires in my heart that I had not been aware of before. One was to be in the ministry, and the other was to write, but again, nothing happened. Life went on as usual with no further leading from the Lord.

Then, about ten years after this, the Lord began to move in my life in a dynamic way. What follows is an account of what He has been doing. It has been a wonderful adventure, and I believe you will enjoy reading about it.

To give you a brief overview, the Lord called me to go to the Republic of the Philippines and minister His Word. As I was preparing to go, He told me not to become associated with any other ministry. After that, He said not to raise financial support, but to trust Him and believe His Word, and He would supply all of my needs.

When I heard this, I was concerned that what I was hearing was not coming from God; but as I spent time with Him in the Word and in prayer, He led me as He had others whom He had called.

For example, He showed me how He had sent out the original twelve apostles: "And He called the twelve to Him, and began to send them out two by two, and gave them power over unclean spirits. He commanded them to take nothing for the journey except a staff - no bag, no bread, no copper in their money belts but to wear sandals, and not to put on two tunics" (Mark 6:7–9).

We also read what Paul said in Galatians 1:1: "Paul, an apostle (not from men nor through man, but through Jesus Christ and God the Father who raised Him from the dead)."

On My Own

Because I was not associated with an organized or established ministry, I had to enter the Philippines on a tourist visa. And because of the restrictions on my visa, I could not do anything to provide financially or materially for myself or for my ministry. I could not take up offerings, work, or do anything else to provide support. In addition, many people in foreign countries think that Americans are rich, so they are not prone to giving us anything.

Because of the Lord's directions, and because I was on a tourist visa, I was totally on my own and without natural support. I found myself completely dependent on Him. What a blessing it was to be free of all care and concern! I could now devote myself to serving Him and others. I could now be fully occupied with following Him and doing what He told me to do. Another benefit was that everything was provided—and on time.

As I have obeyed and followed the Lord, my life has not in any way been less in the quantity or the quality that I'm accustomed to. I have never been late with a payment or paying a financial obligation because of lack of provision on His part.

I was a vice president of a steel company when the Lord called me into the ministry, and I lived better in the Philippines throughout those missionary years than I did here in the United States. Our God is truly awesome! He is great, and I love Him.

He told me that if I trusted in man, I could only go where man could send me and stay where man could keep me, but if I trusted in God, I could go where God could send me and stay where God could keep me. As I have followed His leading, I have been wonderfully blessed. Everything has been His mercy and grace; I could never do enough to deserve or earn what He has done for me. "It is better to trust in the Lord than to put confidence in man" (Psalm 118:8).

Because others have seen what He has done for me, they have been encouraged to believe His Word and trust Him. Not only does He want to supply all our needs, but also He wants our lives to be a testimony to others. "Remember those who rule over you, who have spoken the Word of God to you, whose faith follow, considering the outcome of their conduct" (Hebrews 13:7).

The Normal Christian Life

As you read this account, you will live many of these exciting adventures with Jesus with me. When we live a life of faith, we live the life of God—and a wonderful life it is. The Lord does not want us to be uncomfortable with or afraid of the supernatural. It is supposed to be the normal, natural life of every Christian. "For we walk by faith, not by sight" (2 Corinthians 5:7). "The just shall live by faith" (Romans 1:17b). "But by the grace of God I am what I am, and His grace toward me was not in vain; but I labored more abundantly than they all, yet not I, but the grace of God which was with me" (1 Corinthians 15:10).

I am not encouraging anyone to do anything the way I did. The Lord has His plans and purposes for each of our lives. Each life, each calling, and each ministry is different. "There are diversities of gifts, but the same Spirit. There are differences of ministries, but the same Lord. And there are diversities of activities, but it is the same God who works all in all. But the manifestation of the Spirit is given to each one for the profit of all" (1 Corinthians 12:4–7).

Each person should clearly and distinctly discern the Lord's will and then follow Him as He leads. "For as many as are led by the Spirit of God, these are sons of God" (Romans 8:14).

At the time, I didn't see it, but after the Lord's calling, He began leading me in much the same way He led Abraham. "By faith Abraham obeyed when he was called to go out to the place which he would receive as an inheritance. And he went out, not knowing where he was going" (Hebrews 11:8). Notice how he was called. He did not go out on his own. I knew for years that I was going to the Philippines, but it wasn't until I was actually on my way that the Lord spoke and told me what I was to do when I arrived.

As the Lord led, I prepared. Since I didn't know what He had called me to do, I didn't know what to prepare for, so I prepared myself for Him. "And this they did, not as we had hoped, but first gave themselves to the Lord, and then to us by the will of God" (2 Corinthians 8:5).

As I trusted and followed Him, He prepared and sent me out. This is the account of those wonderful years of walking with Jesus.

CHAPTER 1

Life That Truly Is Life

One evening after I had graduated from high school, as I was contemplating what I should do next, I had a strong impression from the Lord. It seemed to me that I should join the U.S. Air Force.

When I joined, I had the desire to be a radio operator. At that time, when you entered the Air Force, they would not guarantee you any specific job. There were four basic career fields, and all of the jobs were organized and placed in them. You could select your career field but not your job. Since I wanted to be a radio operator, I selected the field it was in, hoping this is what they would give me.

When I received my orders, I happily found that I had been assigned to radio operator's school. One of the reasons this was such a blessing is because during my time in the Air Force, I heard of only one other person getting the job he wanted. Years later, when I began to learn how the Lord leads and provides, I realized that He was the One who had done it.

After radio operator's school, they asked me where I would like to be sent. Since I didn't like hot weather, I asked to be sent to England, Scotland, or Germany, but they sent me to the Philippine islands!

After a year and a half in the Philippines, I was sent to Luke Air Force Base outside Phoenix, Arizona, where I was ultimately discharged. After my discharge, I decided to return to Dallas, Texas, where I had lived with an uncle and his family and had completed high school. I felt that my prospects would be better there.

It was shortly after arriving back in Dallas that I began to sense the desires to be in the ministry and to write. Since I had taken an apartment near Southern Methodist University, I thought about attending school there so I could prepare for the ministry. However, my money was running out, so I took a job and soon forgot about the ministry and writing.

God's Dealings

At that time, I believed the same about the Lord Jesus that I do now. I had never heard about being born again, so I don't know if I was saved or not. However, since I had been brought up in a Christian home, I just assumed I was all right with God. Since I was not walking with the Lord, I was pretty empty. Even though I was doing well professionally, there was always a big void in my life.

God had been dealing with me for years, but I didn't see it at the time. Soon after I got to Dallas, I went to church, but it seemed to me that there was a lot of talk and no action, so I decided church wasn't for me. About five years later, I went to church again. It was the same as before. Of course, I was the one who was wrong, not the churches!

As I think back, I remember that a man and wife, from whom I had rented a room in Phoenix, had invited me to go to church; but I wasn't interested. After I moved to Dallas, I remember how the Lord had sent people to witness to me, but I wouldn't listen.

Once while I was waiting in a doctor's office, I picked up a Bible that was on a table and began to read it. How alive and interesting it was! I remember thinking that I ought to buy a Bible and read it, but that too I soon forgot and went on about my life. How patient, loving, and merciful our God is!

> The Lord is not slack concerning His promise, as some count slackness, but is long suffering

toward us, not willing that any should perish but
that all should come to repentance. (2 Peter 3:9)

After about nine years, the steel fabricating company I was working for decided to set up a branch in Houston, Texas. They made me vice president and transferred me there. It was a good job with wonderful potential.

Life in the Pits

Life should have been great. I had excellent health, a good mind, plenty of money, no bills, and, as far as the president of the company was concerned, I had a job for life. Actually, he was preparing me to take over the company when he retired. I was years ahead of my peers in business, and all without having completed a year of college.

I had all the benefits and perks. I was on top of the world. But I was so miserable that I decided to commit suicide! As far as I was concerned, I had everything life had to offer, but I was not fulfilled or happy.

There was only one more step in business, and that was to own a company. I had already been president of my own company before going to work for this company, so that wasn't anything to look forward to. The top of the world without the Lord is the pits! I had lost interest in life and was seriously considering suicide. I just hadn't decided how I was going to do it.

Then, I finally attended church again. On the second Sunday I went, I heard the Gospel of Jesus Christ for the first time. I'm not saying it was the first time the Gospel was preached to me; I'm saying it was the first time I heard it.

Today this Scripture is fulfilled in your
hearing. (Luke 4:21)

The Word of God is only fulfilled in our lives when we hear it.

That day, I heard the Gospel of Jesus Christ and believed. When the pastor invited anyone who wanted to receive Jesus to pray with him, I prayed. When I prayed, I didn't sense anything happen, but over the next few weeks, I started to change. The *new birth* or my fully committing my life to God began to work its way out as I spent time with Him in church (where I wanted to go now), in Bible reading (I now began to understand the Bible), and in prayer. Thoughts of suicide were gone. The Lord had dealt with that issue.

The amount of business we were doing in the new branch office was phenomenal. The Lord was using that company to accomplish His will in my life and was blessing them for it. Our God blesses and makes us a blessing.

> To you first, God, having raised up His
> Servant Jesus, sent Him to bless you. (Acts 3:26)

The company president was so happy that he would call and say, "Spend more money." What he meant was, "Enjoy yourself. Entertain more. Spend more company money."

Tulsa Beckons

After a few months, it seemed to me that I was going to move to Tulsa, Oklahoma. This was strange because I don't remember ever having been there, I did not have any friends there, nor had I ever done business there. Yet it seemed so real to me that I began telling business associates that I was moving to Tulsa. When they asked what it was like in Tulsa, I'd tell them that I didn't know; I'd never been there. When they asked what I was going to do there, I would also say I didn't know.

A friend told me he would contact a company he was affiliated with, which was about fifty miles from Tulsa in Muskogee, Oklahoma. He said that he would ask if they might have a job for me. They seemed interested, so I took a day off and went to visit them. They were friendly, but I could tell there was nothing there for me.

Then one morning, as I walked into the office, things seemed to change. The peace was gone. I was saved, successful, and happy; but all of a sudden, my contentment was gone. All that the job and the company meant to me was gone. I was uncomfortable there.

I decided I didn't want to be there anymore. After a while, I went into my office, shut the door, and sat down at my desk. I thought, *Maybe I should pray about the situation.*

I prayed and asked the Lord for guidance. As I continued to sit there a while longer and think, the receptionist buzzed and told me that the president of a Tulsa company whom I had previously met was on the line.

When I answered, he asked, "When are you coming up here to work for us?" When I told him I was interested, he asked if I could fly up that day and talk to him. I told him that wouldn't be possible. (I had just received an $880,000 contract on a job that I had sold at the University of Texas, and it would take about two full weeks of all my time to get it ready for production. That was by far the largest job I had ever sold. I now believe that the Lord was ready to move on in His plans for me, and He was repaying them for what they have done for Him and me. God always pays for what He orders.) When I told him I couldn't leave Houston at that time, he asked if he and their vice president flew to Houston that evening, I could meet them at the airport to discuss their offer. I told him yes, and the arrangements were made. Now, I was really starting to believe in prayer!

As I was driving to the airport that evening, I thought of five different conditions I thought were essential for me to go to work for their company. I talked to the Lord about these five things and then forgot about them. That evening as we talked, they brought up every point I had prayed about and responded according to what I had requested from the Lord.

A Counteroffer

About a week after I got to Tulsa, the president of the company I had worked for in Texas called and asked me to go back to work for him. He said that if I did, he would attempt to buy a company in Texas that did business in the United States and abroad, and it would be mine to manage. I told him I would pray about it and let him know. I prayed that night but had no leading from the Lord to return to Texas. I appreciated his offer, but since I believed Tulsa was where God wanted me, I turned it down.

As my daily life continued, unknown to me, the Lord was moving me on in His plans and purposes for my life.

Life That Truly Is Life

In Tulsa, there is a crisis intervention telephone counseling service. I volunteered as a counselor and got to see firsthand the lost, the suffering, and dying people of this world. One night, a woman I was talking to was committing suicide and was dying as I talked to her. The reason she had called was so we would send someone to her house to be there when her children arrived home from the movie she had sent them to. When the emergency people arrived, she was already dead.

Some of the terrible, tragic situations that I came in contact with are almost unbelievable, and I soon realized that without Jesus Christ, there is no hope.

The staff and volunteers worked hard trying to help, but I could see that psychological training and counseling techniques without God and His Word are of little value. Some people were being helped to cope and make it through this life, but it is only through God and what He has provided through Jesus that we can receive health and wholeness. God is light, God is life, and God is love and everything else. It is only through Him that we can receive the new life we need. Only He can change us!

The Only Answer

When I saw how we were failing to help the callers, I began to minister the Lord Jesus Christ to some of them, and wonderful things happened. I began to see in practical application that Jesus Christ is the only answer to every problem and situation in life.

> For in Him dwells all the fullness of the Godhead bodily; and you are complete in Him, who is the Head of all principality and power. (Colossians 2:9–10)

Please don't misunderstand me. Much of what crisis intervention services do is good. They are able to comfort, encourage, and strengthen many. They are also a good source of information for the many public and private services that are available.

The Lord did several things that confirmed to the callers and to me that Jesus is the answer.

One night, I was talking to a man; and the Lord, to get his attention, gave me a vision of him and what he was doing. I could see that he was lying on a bed smoking a cigarette as he talked to me. When I told him this, he was surprised and asked how I knew that. I told him that the Lord had showed me. This opened the door for me to minister the Lord Jesus to him.

Dealing with a Suicide

Another time was when it was much more serious. A man who claimed to be a member of a major world religion called and said that he was going to commit suicide. He said that he had already put his wife and children out of the house and had barricaded the doors. He also told me he had a gun.

We had been trained that one of the first things to do is get the caller's telephone number. That way, if we are in a crisis situation and the caller hangs up, we can call him or her back. As we talked, I got his telephone number, and after a while, he hung up. He was serious about committing suicide, so I called him back.

As we talked through the evening, I explained that he was worshipping a dead god, and that was the reason he was having these thoughts of death. I talked to him about the Lord Jesus and how God had provided good things for us through Him. I kept talking and talking, but couldn't tell if anything was getting through.

After several hours of conversation, I could tell he was done talking, and now, he was ready to kill himself. I knew our conversation had gone as far as it could. I didn't know what to do next.

Because he was having a problem believing in the reality of Jesus Christ, I said to him, "If you want to know if Jesus Christ is real, ask Him to reveal Himself to you. And if He doesn't, pull the trigger!"

I have great confidence in the Lord; however, I had confronted this man, and one thing you should never do is to confront a suicidal person. You may confront alcoholics and people with those kinds of problems, but never a person who is suicidal. He hung up, so I called him right back, but he didn't answer. I was quite concerned and kept calling him, but got no answer. They told us at the hotline to never call for emergency aid unless the caller requested it, so all I could do was pray.

A few days later, he called back and left a message for me. To the best of my recollection, this is what it said: "Tell the young man that I talked to on Wednesday night that what he said was true and that everything is great." I was the only man on duty that night, so it had to be me. That was the only time I ever received a call back from a caller. The Lord knew I needed to hear back on that one. Did I ever rejoice!

Children's Problems

As I continued volunteering on the hotline, I saw something else: The children had no one to counsel them. They called and wanted to talk about their problems and find help for drug abuse, peer pressure, home and family problems, school problems, abortion, and all of the other things that children are confronted with today, many of which my generation had not experienced. None of us adults knew enough about the children's problems to help them, and they knew it. After we had talked for a few minutes, they would realize we were unable to relate to them and their problems, and they would hang up.

It probably was good for them that most of the counselors were unable to relate to them, because had they been able to, they would have counseled them in the ways of the world, which would have brought even more destruction into their lives. Jesus is the only answer!

After a year and a half of counseling at the hotline, the Lord put it into my heart to start a hotline for children. I did and named it Youthline. It was a great success!

To make young people aware of Youthline, I bought advertising on several radio stations. One was a Christian station, and the other was secular, with a format for youth, but that was expensive.

However, the Lord had foreseen this need and had made provision. At the hotline, I had become aware of public service advertising spots on television. I never in any way tried to conceal that we were a Christian organization or that our purpose was to minister Jesus Christ to the callers. Yet the Lord gave me favor with the secular radio station and all three of the network television stations in Tulsa, and they gave us thousands of dollars of free public service advertising each month, much on prime time.

One television station would send us an invoice each month that was stamped "paid," and it was always for more than $3,000. I don't know how much free advertising all of the stations combined were giving each month.

God Provides Counselors

For the first three months, I opened the telephone lines from 8:00 p.m. to midnight, on Friday and Saturday nights only. During this time, I was learning how to do what the Lord wanted done. After this period, I knew it was time to expand Youthline to seven nights a week. But where could I get enough born-again, Spirit-filled, qualified counselors to fill all the shifts?

At the time, I was attending a Bible training center the Lord had directed me to. One day, about 1,600 first-year students were in the auditorium, and the director of the school said that he wanted a few people to give testimonies.

We knew each other because I had been working in their building program. That day, I was sitting at the back of the auditorium, and I didn't see how he could possibly see me. However, I prayed and asked the Lord to have him call me so I could tell the students about Youthline and ask for volunteers. The next thing I knew, he looked right at me and asked me to come forward to speak.

I told them about Youthline and that we needed counselors. Later, after I finished interviewing the volunteers, we had twelve young men and women and two grandmothers who were Spirit-filled and could relate to and minister to the needs of our callers. Some had been delivered from drugs, some had had abortions, and many had had the same problems and had lived the same destructive lifestyles many of our callers were living.

Now, each was gloriously saved and was ready to minister the same victory they had received through Jesus Christ. Not only were they great blessings to Youthline, but also I know this experience helped to prepare them for future ministry.

Special Training

Before the new counselors were put on the line, they were given special training. The evening of the first training session, just before I left home, I prayed and asked the Lord if there was anything special that He wanted to say to them.

He gave me two things for them. He said that as we get closer to His return, the world would get so dark that even unbelievers would be able to see the difference between righteousness and unrighteousness. That was 1979, and we are seeing that happen now.

The second thing He said was that parents and grandparents of children today have had twenty, thirty, or forty years to come to the Lord; but when Jesus comes, there will be 8-year-olds, 9-year-olds, etc. He said that the closer we get to the Lord's return, the more ministry there will be to children. And we are seeing that happen too.

One Saturday evening, before the other counselors were added, a woman who was interested in becoming a counselor and her husband came to visit. She was interested but didn't think that she could do it. We, however, were encouraging her that she could.

As we sat there talking, one of Youthline's telephones rang. I answered it and motioned for her to pick up the monitor that was connected to that line. As I talked to the caller, she could hear our conversation.

The caller was an 18-year-old young man, and as is so often the case, he did not start by telling me his real problem. Because they are so hurt or sensitive to the attitudes of others or so ashamed or confused, many people will not tell you their real problems. They will start out by telling you some problem they have made up to see how you will react.

Opening Up

If you respond in love and sensitivity and they can tell that you are genuinely interested in them and their problems, they will begin to open up and share with you.

I don't remember what the young man started out telling me, but as he talked, the Lord told me that he was a homosexual. By the way, the Lord only tells us what is pertinent to our ministry to a caller. He never tells us anything else personal about anyone.

As he continued to talk, the Lord directed me to 1 Corinthians 6:9–10 that says, "Do you not know that the unrighteous will not inherit the kingdom of God? Do not be deceived. Neither fornicators, nor idolaters, nor adulterers, nor homosexuals, nor sodomites, nor thieves, nor covetous, nor drunkards, nor revilers, nor extortioners will inherit the kingdom of God."

The prospective counselor heard the conversation and saw me turn to that scripture. She knew that nothing had been said or inferred about homosexuality. After he had talked for a while and I knew it was time, I said, "You are a homosexual," and he agreed.

I asked him if he had a Bible, and when he said yes, I asked him to get it and turn to that scripture. When he had done this, I asked him to read it. He read it and said in a shocked manner, "Do you mean I can't go to heaven?"

I ministered to him and prayed with him. I knew he would need more ministry and counseling, so I referred him to my pastor who has the gift of discerning of spirits. When it was all over, the visitor saw clearly that we minister by the anointing of the Holy Spirit—that it is He who works the gifts in us and ministers through us. And as we trust in and rely on Him, we can do whatever He calls us to do.

A Good Report

We seldom hear about what happened to those we counseled, but every once in a while, we did. After my pastor had counseled him, with his parent's consent, it was decided that he should go to a Christian live-in counseling center for former homosexuals in northern California.

After he had been there a few weeks, the counseling center contacted this young man's parents to see if he had shown up back in Tulsa. They said that he and some of the other young men had left the center, and they could not find them.

They left in one of the young men's Volkswagen. A few weeks later, they returned. They had sold the Volkswagen and bought a van. In the van, they had a number of other young men. They had gone to San Francisco and ministered to other young homosexuals on the streets! Then, they loaded them in the van and took them back to the home.

When the people at the center reprimanded them for leaving like that, they responded, "We couldn't stay here like this when there are others out there who need help!"

Those who were in charge of the home said it was unusual. They said that while those young men were gone, some of the other people who were being counseled in the home all of a sudden became ready to leave, so when they got back with the new people, there was room for them.

By the way, this young man told me that a Christian girl who lived next door to him had been ministering to him and giving him Christian tapes. Don't give up on those to whom you are ministering!

Before I started Youthline, the Lord gave me a strong witness in my spirit that I would be going to the Republic of the Philippines. After Youthline was in operation for about nine months and things were going well, I thought about what would happen to Youthline when I left. I knew that many of the young people who were calling had been rejected all of their lives, and the last thing they needed was for us who proclaimed the Lord Jesus Christ not to be there when they called.

Because I thought I would be leaving for the Philippines as soon as Bible school was out, I decided to close down Youthline. I thought the best way to close was to shut down all of the advertising right away and keep the telephones open for another few months. That way, we wouldn't attract any new callers, and the current ones would not be calling any longer. I didn't want anyone to call and no one be there.

I called all the stations that were carrying our advertising and told them we wanted to cancel all advertising immediately. When I told the counselors, they were disappointed, but I didn't know what else to do.

Hearing from God

Most mornings before school, I would spend time praying for Youthline. The morning after I stopped the advertising, as I sat praying, the Lord spoke to me in an audible voice and said, "Give Youthline to Robin Anderson."

Robin was perfect for the job. She had been one of our counselors from the time we expanded, and she was dedicated and sincere. She loved the Lord and the callers.

Robin was an excellent choice, and she had a degree in psychology. What made it even better was that her parents had a 24-hour a day, seven-day a week Christian hotline in their home in Colorado, so she knew the great responsibility it involved and the restrictions it would place on her life. God had again seen ahead and provided!

I was excited about hearing the Lord's voice, but I think I was more excited about Youthline being saved. Sometimes, we must rearrange our priorities.

Once, as I was praying in the Philippines, the Lord spoke to me and said, "One of Satan's sins was that he wanted all I have. He just didn't want Me."

I knew as He spoke that I was guilty of the same thing to some degree! I didn't seem nearly as excited about the Lord as I was about the ministry. It was time for me to repent of that attitude. Sometimes, we let our lives and our ministries get to be so big and important to us that we cease to be His servants and we try to make Him ours.

Delivered from Cigarettes

Another time the Lord spoke to me in an audible voice was two years before, in 1978. I had been smoking for twenty-two years and couldn't quit. I had tried every plan and scheme that I could think of, but nothing worked. Cigarettes at that time were 50 cents a pack, and since I was smoking a pack a day, I was spending about $15 a month. At that time, I was supporting children overseas for $15 a month, and I began talking to the Lord about it.

I said, "Lord, this is ridiculous. I'm spending $15 a month on cigarettes, and with that $15, I could be supporting a child overseas. I could give him clothes, teach him about Jesus, send him to school, and provide for his needs."

I was smoking a cigarette at the time, and a voice spoke to me out loud from the passenger's side of the car. It said, "If you want to quit smoking, put it out." I put the cigarette out in the ashtray. I didn't feel anything different, but I haven't smoked another cigarette since, and it has been more than twenty years.

In my coat, I had a pack and a half of cigarettes, and at the office, I had a pack in my desk. I couldn't believe it—I didn't have the craving to smoke anymore. It was as if I had never smoked. However, I didn't want to say anything to anyone else in case it wasn't true. Then, after three days, I decided that it was, and I told others what the Lord had done for me.

In the years since, the thought has come to me a few times, *Wouldn't you like to have a cigarette?* I think, *No,* and the thought leaves. Since that day, I have never had a craving or desire to smoke.

Reversing Course

After the Lord told me to give Youthline to Robin, I called to tell her what the Lord said, but no one answered. That was strange. Robin shared an apartment with two other young women students, and whenever I would call early in the morning before, I always got someone. So I decided that when I saw Robin at school, I would talk to her. When I looked for her and couldn't find her, I started to wonder what was going on. I tried to call her again after school, and I still didn't get an answer, so I thought, *If it's not the Lord's will for me to talk to her, I'd better let Him do it.*

The next day as we were changing classes, Robin came up behind me and said, "Tom, I want to talk to you. I want to take Youthline. I called my parents last night and talked to them about it, and I think I should do it." Then, I saw that the Lord didn't want me to talk to her about it. This was something He had to do.

Since then, I have not told anyone what the Lord has told me about a situation that involves them. What they do is between Him and them, and I believe He should be the One who leads them. Besides that, what they do should be in obedience to Him.

That afternoon, I called all the stations and told them we wanted to continue the advertising, and they all put our advertising back on. Youthline continued to operate for about four more years.

New Opportunities

When I moved from Houston to Tulsa, my new employer said that they would pay all of my moving expenses. For some reason, that just didn't seem right to me. It seemed it would be fair to let them pay half, but no more. Later, I found out why. After about seven months, I knew it was time for me to leave that company. It would have been unfair to let them pay that much money for that short of a stay.

Just before I left the company in Texas, they decided to discontinue their profit-sharing plan. Because of that, I had the extra money to pay my half of the moving expenses to Tulsa, money I would need to get set up in Tulsa, money to pay extra taxes, and a $2,000 down payment on a new car.

About the time I was thinking about leaving the company in Tulsa, the Lord directed me to a new church. At the new church, a man offered me a job. It was a construction company, and I knew nothing about construction. He said that he wanted me to be a liaison between him and the various job sites. The salary was 75 % of what the other job paid, but I thought I should take it. At that company, I began to learn about construction.

After about eight months with that company, I knew it was again time to leave. I didn't have another job to go to, so I began to attend a week of Christian meetings. On the Saturday of that week I learned Mark 11:24.

Mark 11:24 Makes the Difference

> Therefore I say to you, whatever things you
> ask when you pray, believe that you receive them,
> and you will have them. (Mark 11:24)

On Saturday afternoon, my brother came to the meeting and told me my first foster daughter, Rhonda, had called and said that Tara, her two-year-old-daughter, was in the hospital, having been diagnosed with spinal meningitis. At that time, 1978, it was incurable. I left immediately and drove to Dallas.

When I got there, we prayed for her according to Mark 11:24. When they ran the next test, they thought someone had made a mistake because nothing showed up. They ran another test, and again, nothing showed up. The doctors were so amazed they asked Tara's parents if they could bring this case up for discussion at the monthly hospital meeting. We should have told them that we had prayed for her, but I didn't think about it. Later, when someone would ask Tara what happened, she would say, "Jesus healed me."

Tara stayed in the hospital a few more days because after she was healed of meningitis, she came down with pneumonia. In that few days, she recovered and went home. I learned from that situation to remain vigilant after a victory.

Remember Paul in Acts chapters 27 and 28? After the storm and shipwreck, when everything seemed safe and secure, the serpent came out of the fire and fastened on his hand.

More Provision

Upon returning to Tulsa, I started to look for a job, but I couldn't find one anywhere. I even called the man I had worked for in Texas before coming to Tulsa, and he wasn't a bit interested. I was running out of money and needed to get a job. When I was at the meeting, I bought some teaching tapes, one of which was blank. I didn't think much about it and put it aside. At the meeting, I had also ordered some tapes to be mailed to me, and when I received them, I found that one of them was blank.

Since I had two blank tapes, I thought I would take them to the ministry and get them replaced. When I got there, I saw a big construction site. I went into the ministry office, and after they had replaced my tapes, I began to think about that big construction job outside. I now had construction experience, and I had a resume in the car, so I asked the receptionist if I could apply for a job. She told me I would have to go to the construction office to apply. When I went there, they took my application and resume, and I left.

The next day, they called and asked me to come for an interview. During the interview, things went well, but the man who interviewed me was noncommittal, so I left not knowing if I would be hired or not.

Later that day, I began to think about a promise I had made to Rhonda and Murry, her husband. They had just bought a new home and had mentioned they wanted to put up a fence around the backyard so the children could go outside and play. I asked them how much it would cost, and they told me about $300.

I knew they didn't have the money, so I told them that I would pay for it. I was going to send the money to them, but since I was having a difficult time finding a job, I thought that I would wait for a while.

My Real Source

But as I thought about that promise, I realized that if I really believed the Lord supplies all of my need, I shouldn't delay sending them the money. That money was not my source, the Lord is. So I sat down and wrote the check and mailed it. That left me with a little over $10 in the bank. I thank God for the opportunities He gives us to exercise faith so He can bless us!

The next morning, the man who had interviewed me called and asked me to come back to talk to him. As we sat there talking, he said that he would hire me and would pay me seven dollars an hour. I thought this meant I would be paid $280 a week. It was hard for me to remember when I had last made only seven dollars an hour. However, the Word that was hidden in my heart came forth. Without even thinking about it, I said, "My God shall supply all of my need according to His riches in glory by Christ Jesus. I'll take the job."

The job that was available was in supervision, and the construction job I had just left had trained me in construction supervision.

The project manager knew I had steel fabrication experience. That is what I had done in Texas and had come to Tulsa to do. A few days after starting my new job, he asked if I could make him a list of the steel items on one of the new buildings they were getting ready to build. When I told him I could, he asked me if I could do it that weekend, and I said yes.

As we parted, he said to be sure to keep track of my time and turn it in, and he would pay me. For years, I had worked on a salary, and no matter how many hours I worked, I was always paid the same. I told him he didn't need to pay me; I would be glad to do it without pay, but he insisted, saying that he also wanted me to keep a record of all my time and to submit it each week.

Because of that, and because of the other overtime hours they gave me and the raises that came quickly, from the first week on, I made more money there than I had ever made on any other job! The Lord is certainly faithful to fulfill what He has spoken to us.

> You have seen well, for I am ready to perform My word. (Jeremiah 1:12)

More Leading

I took this job in August 1978, and everything moved along smoothly as the Lord provided wisdom, understanding, and ability to do a good job. Then, in the fall of that year, I began to sense in my spirit the Lord's leading to go to the Philippines. It would be almost three years before I left, but I can see now why He told me so far in advance. The reasons were so I could begin to prepare spiritually and get out of debt.

The next spring, the Lord told me to go to school, so I enrolled for the term that would begin that fall. At the time I was to start school, another opportunity presented itself. There was a really good steel fabricating company I knew about in San Angelo, Texas. I had wanted to go to work for that company for years, but there had never been an opportunity. Then, at about seven o'clock in the morning the day I was to start Bible school, I received a telephone call from a salesman I had known in Texas. He told me that this company now wanted me to work for them. I turned them down.

As far as I was concerned, this was a temptation from the enemy to divert me from the Lord's will and plan. I had heard from the Lord, and I knew that going to school was His will.

When I discussed my going to school with my employer, they agreed to let me work afternoons only, since the Bible school classes were conducted in the morning.

Through the summer, I set up my finances so I could live on the money I would make on my job. When I thought I was all set, I had an opportunity to have my faith tested.

First, the place where I was working would not pay me for my Labor Day holiday, which I thought I was due. Not only was it a test of my faith for provision, but also it was a test to stay in love and forgive. Many tests and trials seem to come when you are in transition and not on familiar ground or after great victories when your guard is down.

The Next Trial

By God's grace, I made it through that test, and then, the next trial came. A few days later, I was informed that I would no longer be needed on my job. I wanted to work afternoons so I could study evenings. You can imagine how hard it would be to find an afternoon-only job that paid as much as I was making and needed to make to support our home and pay tuition and school expenses. Praise God, I believe He supplies all our needs, so I continued in school and trusted Him.

Later that week, the place where I had been working called me. They said they were starting another work crew and asked if I would come back and supervise it at the same pay rate that I had been making. They also said they understood I could only work afternoons.

> Let us hold fast the confession of our hope
> without wavering, for He who promised is faithful. (Hebrews 10:23)

Everything went well until the first of the year. In early January, my employer terminated me. Because my job was outdoors, we couldn't work any longer because of the weather. For the next two and a half months, I couldn't find a job anywhere doing anything! Because I had always paid our bills on time and had good credit, the creditors were patient.

The real problem was the school I was attending was a one-year school. At the end of the first year, you graduated and left if you wanted to. However, they also offered a second year. When the Lord told me to go to school, I thought it was for one year. That's not what He said; it's what I heard.

God had not failed to make provision. During the autumn, my employer had asked me if I would take over a new position, which was managing the new campus maintenance department they were forming. Since I thought I would be going to the Philippines after graduation, I turned them down.

Missing God

Thank God, I continued in school. He provided the grace that I needed not to get discouraged and to fully apply myself. I only missed days in school when I didn't have money for gasoline.

Not realizing that I had missed the Lord's provision, I continued where He had put me and did what He had given me to do as I continued to look to Him. I honestly don't know how we made it. But one thing we did: We always shared and gave. Since I was not earning any money, we had no tithe. Because we wanted to give, when we went to church, we would search the house, looking in the furniture and in the pockets of clothes hanging in the closets for money for offerings; and we always found some.

> In this manner, therefore pray: Our Father in heaven, Hallowed be Your Name.
> Give us this day our daily bread.
> (Matthew 6:9,11)

I believe part of what He was teaching us is that our Father has only promised to give us what we need on the day we need it.

One day, I saw that He created us to live in a garden. When you live in a garden, what you need is always right there around you. All you have to do is reach out and pick it. The way we pick what God has provided for us is to reach out by faith.

I didn't ask anyone for anything or borrow anything, even though it is not wrong if a person does. It never occurred to me. I was so convinced that God was going to supply our needs, I kept waiting for Him.

Treasures of Darkness

> I will give you the treasures of darkness.
> (Isaiah 45:3)

In some of the darkest times of our lives, the Lord reveals and gives great treasures to us.

One of the things I treasure out of that situation was the time two of my fellow students invited us to Sunday lunch. They are a husband and wife with two children.

He and I had been working together, so when my job was terminated, so was his. I knew they were in the same financial situation we were, just making it day to day with nothing to spare. However, when we sat down to lunch, they served us steak. I couldn't believe it! I saw what a wonderful thing they were doing for us. The cost of that meal was much more than money.

It reminded me of David, when the three mighty men went into the garrison of the Philistines, to the well of Bethlehem, to get David a drink of water.

> David was then in the stronghold, and the garrison of the Philistines was then in Bethlehem.
>
> And David said with longing, "Oh, that someone would give me a drink of the water from the well of Bethlehem, which is by the gate!"
>
> So the three mighty men broke through the camp of the Philistines, drew water from the well of Bethlehem that was by the gate, and took it, and brought it to David. Nevertheless he would not drink it, but poured it out before the Lord.
>
> And he said, "Far be it from me, O Lord, that I should do this! Is this not the blood of the men who went in jeopardy of their lives?" Therefore he would not drink it. These things were done by the three mighty men. (2 Samuel 23:14–17)

How wonderful it is to experience the love of God as others sacrifice themselves for us!

Our Financial Drought Ends

Another thing I can't forget happened near the end of the two-and-a-half month financial drought.

One evening, one of our grandmother counselors arrived early for her shift. There was an exercycle in the counseling room. It belonged to our family, and she asked what we were going to do with it when we left for the Philippines. When I told her we would probably sell it, she asked how much we would want for it. I gave her a price, and she said she wanted to buy it and asked if she could leave it there until school was out.

She took out her checkbook and started to write a check, but I told her that she didn't have to pay for it until she picked it up in two and a half months. She said no, she wanted to pay for it then. At that time, we had no money at all! The only thing we had in the whole house for dinner was a two-pound bag of rice. At that time in my life, having rice to eat was like not having anything to eat. After she gave me the check, I walked quietly out of the counseling room and exploded.

Through the years, the Lord had blessed us with many nice things, but I had never thought about selling any of them to meet our financial needs.

When our finances began to flow again within a period of one week, we were completely out of debt and had enough money left to go to Dallas on our spring break from school. I had prayed and asked the Lord for the money to go there because I wanted to witness to some friends.

Two things happened. The first was, one afternoon, the Lord spoke to me and said to call the man for whom I had worked when I first arrived in Tulsa. When I called and asked him for a job, he gave me a perfect job for my situation.

He started me at a high rate of pay; I could do the work at home; I could work whatever hours were convenient (which meant I could go to Dallas on spring break); and I could work as many hours as I wanted. I think I worked forty-eight hours that first week.

The other thing that happened was when my mother decided to give me a number of guns an uncle of mine had left to her when he died. When I sold the guns and got my first week's paycheck, our financial crisis was over.

CHAPTER 2

Called and Equipped

As the school year progressed, I wondered what my ministry would be. One day, I heard that a group that worked in the Philippines was going to hold a meeting at our school. Their ministry was building churches, and they wanted to see if anyone from our school was interested in working with them. Since I knew I was called to the Philippines and since I had construction experience, I thought I would attend the meeting.

The representatives explained their ministry and the mechanics of being associated with them. However, there was no desire in my heart to build buildings. The desire of my heart was evangelism. I thought, *If my ministry was building churches, I would build churches during the day and hold evangelistic meetings at night. That way, when the building was finished, there would be a congregation ready.*

This was not God's plan for me; it was the apostolic call on my life showing through, which is evangelizing an area and starting a work where no ministry exists. The more I thought about working for that ministry, the less it appealed to me. Not that there was anything wrong with their ministry, but I just wasn't comfortable with the thought of working with them. What seems obvious circumstantially does not necessarily mean it is from God.

In addition, it seemed to me that the Lord was saying when I went to the Philippines, I was not to raise financial support, but to go to the field and trust Him to provide for my needs. That really disturbed me because I'd never heard of such a thing before. However, later, I read about others who had done that. And as I studied it out in the scriptures, I saw that the Lord had sent out the apostles and the seventy that way.

Because I was called to minister abroad, I thought I should get whatever information I could, so I set up an appointment with the head of the missions department at school. He said that he thought the Lord was telling me not to become associated with any other ministry but to go out on my own. When he said that, I had a powerful witness in my spirit that what he said was from the Lord. However, I didn't know you could do that. I thought you had to go out under another ministry.

He also said that before I went, I should itinerate for about a year and raise support. I got no witness on that at all. I asked him how I could get into the Philippines to minister if I did go out on my own. He said that he didn't know, but he thought I had to have a letter of invitation from a Filipino ministry to come into their country to minister. He said that he had the telephone number of a woman in the Philippines, and I could call and ask her.

A Welcome Invitation

When I called her, she said that she was unaware of a letter of invitation being required, but she would write one in case I needed it. As we were finishing our conversation, she asked where I would be staying, and I told her that I didn't know. When I said that, she invited me to stay with her family when I arrived. She added, "We will give you the best room in our house."

Praise God, I had an invitation into the country if I needed it (I found out later that I didn't) and a place to stay with a family. This thing was starting to work out already. The Lord was performing His Word!

Then, I had my first Ishmael. Ishmael was the offspring of Abraham through the flesh after God had given him a child through promise. I didn't know anything about waiting on God's timing. I thought that when school was out, I should leave.

I had already sold our house the previous fall. The man was buying it for rental property, so we were able to keep it until school was out. In May, I started selling all our household items. It's really sad to look back at that time. Because nothing would sell, I practically gave everything away. I had (and still have) a lot to learn about God and His ways. He has ways of fulfilling His plans for our lives and ministries. When it is His plan, in His time, it's good and enjoyable. That way it is done in the Spirit in a godly way. He is glorified, and we are blessed.

Running Ahead of God

When I tried to sell the furniture, I couldn't. Nobody wanted anything. I had cut the prices to practically nothing, but I still couldn't sell it. We sold some of the things that were special to friends because I cared who had them. But other than that, except for our large dining room set, which we wouldn't have been able to use in the apartment that we rented later, I couldn't sell anything.

On that last frantic day when my mother, my son, and I were packing, cleaning, and moving out of the house so I could leave for the Philippines, I was really trying to sell that furniture. Finally, that afternoon, I decided to give it away. Two of my fellow students had gotten married recently, and I knew they could use some furniture, so I started calling them. I kept calling and calling, but no one answered.

By the time I realized I was not going to be able to contact them, it was too late to try to find someone else to give it to. I had to put the furniture into storage. It was amazing because all the furniture I could not get rid of was exactly what we needed to furnish our apartment. The Lord cares about the home we and our families live in.

About a year later, I finally did sell the furniture. Then, we really didn't need it any longer. I'm learning that all we do is by God's mercy and grace through faith.

> He has made everything beautiful in its time. (Ecclesiastes 3:11)

When we run ahead of God, we create problems for Him, for others, and for ourselves.

> But imitate those who through faith and patience inherit the promises. (Hebrews 6:12)

Missing Out

That summer, my employer started a night shift and asked if I would supervise it. This was the second provision the Lord made for me to go to school the second year. I can see now what a great benefit it would have been. I still didn't pick up on the fact that the Lord wanted me to go the second year.

I spent the next year in Tulsa working nights with nothing to do during the day. We must be sensitive to the Lord's leading. Everything He does is for our benefit.

> Beloved, I pray that you may prosper in all things and be in health, just as your soul prospers. (John 2:3)

> Let the Lord be magnified, Who has pleasure in the prosperity of His servant. (Psalm 35:27)

And if He has pleasure in the prosperity of His servants, how much more pleasure does He have in the prosperity of His children!

I was also learning outside of school the benefits of ministering to others. One snowy and icy morning, I saw an older man fall in the parking lot. He was taken away in an ambulance, and I heard later that his back was badly hurt.

During those days, we were having a Bible study in our home. When I got home that night, I was sick with the flu, and as I sat there waiting for the people to arrive, I remember thinking how I wished the Bible study was over so I could go to bed. I had been so busy that day I hadn't even thought about claiming my own healing.

As we prayed that evening for the needs of the group, I thought about that man, so I asked everyone to lay hands on me and pray for him. As they laid hands on me and prayed, it felt like someone had poured warm honey on my head, and it ran down over my body. I was instantly healed. When I saw the man and his wife sometime later, his back had been completely healed.

Learning about Healing

The Lord was teaching me to trust and believe Him for healing. One afternoon, I was again sick with the flu. While I was in a drug store, I looked to see if I could find anything to help me get well. Then, I thought, *No, I'll trust God for my healing.* As I stepped out of the store, I was instantly healed! All symptoms were gone, and I felt great.

Another time was when I was sick with the flu. I was really working hard to receive my healing. I was reading my Bible, praying, confessing the Word, and claiming my healing, but nothing was happening. There is nothing wrong with these things, but I was working and not receiving. One morning during this time, I was sitting on my bed reading the Bible and praying. As I sat there working hard trying to be healed, the Lord spoke to me and said, "Just receive." So I fell back on the bed and relaxed and said, "I receive." I was instantly healed as I rested in His Word.

I was struggling with the idea of going to the Philippines without financial support, but I learned that as we look to Jesus, He provides the faith we need to receive the grace to do His will.

> Looking unto Jesus, the author and finisher
> of our faith. (Hebrews 12:2)

> From there they sailed to Antioch, where they had been commended to the grace of God for the work which they had completed.
> (Acts 14:26)

Through God's grace, we will complete the work He has called us to do.

Now, I would like to share two of the miracles the Lord did to help prepare me to follow His leading.

The first was a physical healing. One Sunday evening, Tom, my son, and I were working out with weights, I pulled my arm out of the socket. Through another incident a few days later, I pulled it out even further. My arm hung at my side, completely useless. I couldn't move it at all. When I sat down, I would pick my left hand up with my right hand and place it on my lap.

The evening it happened, I claimed my healing by the stripes of Jesus. However, I wasn't sure that Jesus's stripes covered bones that are out of joint. I never heard of anyone having a bone relocated.

> And all My bones are out of joint.
> (Psalm 22:14)

Because we were crucified with Him and raised up with Him, our bones were relocated, in Him, at the resurrection; but I didn't know that then. The Lord is mighty in working in us even though we may not know everything He has provided.

Because it was my left arm, I could still work. As the days went by, however, there was no sign of healing. Actually, my arm was getting worse! Day by day, there was less and less feeling in it, and as the numbness increased, so did my concern. I didn't know what was happening in my arm, whether the blood was being shut off or what. I started having thoughts of blood poisoning, gangrene, and losing my arm. As I continued to pray, I said, "Lord, I want to believe what You have promised me for my healing, but if my healing doesn't manifest itself in a few days, maybe I'd better go to the doctor."

Because the bone was out of its socket and the end was on the outside of my shoulder, I couldn't lie on my stomach to sleep anymore. I couldn't lie on my right side because the weight of the arm pulling down on the joint hurt too much. I couldn't lie on my left side either. If I did, I would be lying on that shoulder. All I could do was lie on my back, and that is not a comfortable way for me to sleep.

Miracles from God

As time went on, I became more and more exhausted. On Friday evening when I got home, I was really tired and discouraged. I was still believing and confessing my healing, but there was no change. (I believe this would be a good place to explain what I mean when I say that I was confessing my healing. The New Testament was translated from texts that were written in Greek. In many places, the Greek word "homologeo" was translated as "confession." "Homo" means "the same" and "logeo" means "to speak." God tells us in His Word that He wants us to say the same thing He does about everything. Being a Christian means that I am in covenant relationship with God. That means I now have what God has promised me. Even though my arm was dislocated, I had a promise from God for my healing, and I should be believing and saying what the Lord says about my situation. Besides that, when I meditate on the Lord and His goodness and what He has promised, He is magnified and not the devil or my problems. I am drawing closer to Him: I'm taking His side in the matter, and He is the One who is glorified.)

After dinner, I went to bed. As I lay there in the dark, on my back, I suddenly felt pressure on my shoulder. I could feel that pressure slowly pushing my arm bone back into the socket. I could feel the bones grating together, but there was no pain. I don't know how long it took, probably less than a minute, and then, it was finished. I jumped out of the bed and began to move my arm all around. It was totally healed! Our God is so great!

The second miracle the Lord performed to reveal His power and ability to provide for my needs happened when I was working as the night foreman at the steel company. As the night progressed, it became evident there were not enough beams on the transfer tables to last the whole shift. This was serious because we could not stop production when we ran out of beams. Every hour of lost production would cost the company hundreds of dollars.

That night was raining hard. The steel we would have to use to finish the shift was outside in a railroad car. There were a number of reasons why I didn't want to take men outside to unload the railroad car and to reload the transfer tables in the rain.

It was very dangerous. At two other steel companies where I had worked, men were killed. We would be working down inside an open railroad car with no light at all. We would be hooking our crane to big bundles of steel that weighed up to ten tons, and as we moved them up through the steelyard, men would have to hold on to them to guide them. If a chain was not hooked up properly and came loose, someone could be killed. Besides that, the light in the steelyard was so dim in the rain it would be difficult to see. The men would have to climb up and down over piles of steel lying in the yard as they held on to the bundles of steel to guide them. Also, the crane was powered by 440 volts of electricity; and in that rain, if there had been a short circuit in the electrical system, it could have killed someone. Besides all that, we did not have enough rain gear to go around, so some of the men would be soaking wet.

As the night progressed, I looked out the door from time to time to see how our steel supply was holding out and to see if the rain had let up. Finally, at about four o'clock in the morning, our steel supply was almost gone. As I stood looking at the rain, out of my spirit came these words: "Rain, I command you to stop in the Name of Jesus!" I didn't even think about it; it just came out of my spirit. As soon as I said "Jesus," the rain completely stopped. It stopped instantly. Not another drop fell.

As I stood there amazed, looking at what had happened, the Lord spoke to me and said, "When do you think the rain stopped?" I didn't know how to respond; I was so amazed. Then, He said, "That rain stopped about a minute ago. Those clouds are thousands of feet in the air." Then, I realized the rain had stopped falling almost a minute before I had prayed. My prayer had only lasted a few seconds, but the rain had quit hitting the ground as soon as I said, "Jesus." Then, the Lord spoke again,

> Your Father knows the things you have need of before you ask Him. (Matthew 6:8)

Some of the men and I went outside and safely unloaded the railroad car. The Lord knows and is concerned about all our needs, and He wants to supply them. As He proved that night, He is well able to do it. After those experiences, I was excited and ready to go to the mission field.

Living in a Dangerous Place

I also started to think about personal safety in the Philippines. When I had been there in the Air Force, it was a dangerous place. But I remembered how unhappy the Lord had been with the Israelites when He wanted to lead them into the Promised Land, and they had murmured and said, "Why has the Lord brought us to this land to fall by the sword, that our wives and children should become victims? Would it not be better for us to return to Egypt?" (Numbers 14:3).

He said to me, "What kind of God do you think I am that I would take you and your family into a place where you would be destroyed?" By that time I was convinced that the Lord would supply our needs if I obeyed and followed Him, I was ready to go and trust the Lord for anything and everything.

If I was concerned about the welfare of those men when I was taking them to unload steel from a railroad car, how much more is the Lord concerned about us and our families' welfare as He takes us out?

A few months before I left for the Philippines, an elderly aunt came to visit. She had heard about what I was going to do and flew from Pennsylvania to talk me out of it. She was upset to hear I was quitting my job, giving up my career, and moving to a foreign country. When she asked how we were going to live, I told her that God was going to supply all our need. Of course, I had heard from God and she hadn't, so I could understand her concern.

We need to be careful that our natural concern for someone's well-being does not attempt to move them out of God's will. Do you remember how the Lord rebuked Peter in Matthew 16:21–23?

> From that time Jesus began to show to His disciples that He must go to Jerusalem and suffer many things from the elders and chief priests and scribes and be killed and be raised the third day.
>
> Then Peter took Him aside and began to rebuke Him, saying, "Far be it from You, Lord; this shall not happen to You!"
>
> But He turned and said to Peter, "Get behind Me, Satan! You are an offense to Me, for you are not mindful of the things of God, but the things of men."

Blessings and Challenges

However, the Lord turned my aunt's visit into a blessing for everyone. That Sunday when we went to church, I was greatly blessed to watch her pray to receive Jesus and be born again. As we walk in obedience, forgiveness, and love, we give the Lord opportunities to be glorified and bless other people.

Then, I had another challenge from the devil. A Christian man who is dear to me bought a number of books for me that teach about various world religions. These religions are of the Antichrist for they deny what God says about Jesus and what Jesus says about Himself.

> When Jesus came into the region of Caesarea Philippi, He asked His disciples, "Who do men say that I, the Son of Man, am?"

> So they said, "Some say John the Baptist, some Elijah, and others Jeremiah or one of the prophets."
>
> He said to them, "But who do you say that I am?"
>
> Simon Peter answered and said, "You are the Christ, the Son of the living God."
>
> Jesus answered and said to him, "Blessed are you, Simon Bar-Jonah, for flesh and blood has not revealed this to you, but My Father who is in heaven."
>
> And I also say to you that you are Peter, and on this rock, I will build My church, and the gates of Hades shall not prevail against it. (Matthew 16:13–18)

This man said that I should read the books. He said that as I learned about these religions, I could minister to the people caught up in these deceptions more effectively, because I would understand better what they believed.

As I was reading one night, it suddenly felt as if my mind was being pulled out of my head. When that happened, I destroyed the books. Those are demonic religions, and when you open your mind to them, you are opening your mind to demons.

> For I determined not to know anything among you except Jesus Christ and Him crucified. (1 Corinthians 2:2)

All we need to know to preach the Gospel to every person is Jesus Christ and His crucifixion. We don't need anyone or anything else!

> For I am not ashamed of the gospel of Christ, for it is the power of God to salvation for

> everyone who believes, for the Jew first and also for the Greek.
>
> For in it the righteousness of God is revealed from faith to faith; as it is written, "The just shall live by faith." (Romans 1:16–17)

> But as it is written: "To whom He was not announced, they shall see; and those who have not heard shall understand." (Romans 15:21)

God will cause each person to see and understand what He has provided for us through Jesus. God reveals His salvation by the Holy Spirit through the Gospel of Christ.

That man loves the Lord and me. When he bought those books, he thought he was helping, but God and His Word are the only ones we can receive from and always be safe.

Wrong Advice

Others influenced me to do things that are contrary to God's Word. I was told that when you are in a foreign country, you will be ministering to poor people; and since you don't want to offend anyone, you have to live like they do, eat like they do, and dress like they do. This is not true.

Wherever we are, if we are walking in love and are truly allowing the Lord Jesus to live through us, we will not hurt or offend those who are sincere, even though we may have some things they don't have.

> And when James, Cephas, and John, who seemed to be pillars, perceived the grace that had been given to me, they gave me and Barnabas the right hand of fellowship, that we should go to the Gentiles and they to the circumcised. (Galatians 2:9)

Those who said I would have to live in poverty were not speaking from the Lord by His grace. The Lord provides the grace to minister. If what is being spoken to us is from the Lord, we should be able to perceive His grace. The Lord told me to always look for His grace.

It is possible for Christians to be wrong! Galatians 2:11 says, "Now when Peter had come to Antioch, I withstood him to his face, because he was to be blamed." Not only do we have to constantly test what is being ministered to us by others, but also we must be sure that we are right in what we say and do. The ultimate responsibility to receive what is right and to say and do what is right lies with us.

> Be diligent to present yourself approved to God, a worker who does not need to be ashamed, rightly dividing the word of truth. (2 Timothy 2:15)

I am not opposed to living in poverty if that is what we need to do to be able to minister to poor people. However, the Lord has called us to help the poor, not to live and suffer like them. Of course, we must use wisdom and discretion. Once when we took a group of wealthy women to a squatter area to minister, we told them to take off most of their jewelry and dress neatly but plainly.

Blessings and Abundance

> But when you pray, do not use vain repetitions as the heathen do. For they think that they will be heard for their many words.
>
> Therefore do not be like them. For your Father knows the things you have need of before you ask Him.
>
> In this manner, therefore, pray: Our Father in heaven, Hallowed be Your name.

> Your Kingdom come. Your will be done on earth as it is in heaven. (Matthew 6:7–10)

God does not want us to have to wait until we get to heaven to have what He has provided for us. He wants us to have it now!

When Jesus fed the 5,000 men plus women and children, the Bible says, "So they all ate and were filled" (Matthew 14:20). God wants His people to be full! This is borne out in 2 Corinthians 9:8:

> And God is able to make all grace abound toward you, that you, always having all sufficiency in all things, may have an abundance for every good work.

> The Lord is my Shepherd; I shall not want. (Psalm 23:1)

> Oh, fear the Lord, You His saints! There is no want to those who fear Him.
> The young lions lack and suffer hunger; But those who seek the Lord shall not lack any good thing. (Psalm 34:9–10)

> Praise the Lord! Blessed is the man who fears the Lord, Who delights greatly in His commandments. His descendants will be mighty on earth; The generation of the upright will be blessed. Wealth and riches will be in his house, And his righteousness endures forever. (Psalm 112:1–3)

God's Word is full of great and precious promises that are available to whosoever will call upon Him.

> For there is no distinction between Jew and Greek, for the same Lord over all is rich to all who call upon Him. (Romans 10:12)

> Grace and peace be multiplied to you in the knowledge of God and of Jesus our Lord, as His divine power has given to us all things that pertain to life and godliness, through the knowledge of Him who called us by glory and virtue, by which have been given to us exceedingly great and precious promises, that through these you may be partakers of the divine nature, having escaped the corruption that is in the world through lust. (2 Peter 1:2–4)

> I have come that they may have life, and that they may have it more abundantly. (John 10:10)

The Gospel is good news: it is salvation.

Since it has been the Lord who has provided, my lifestyle has been His will, not my own. Overall, I have lived better in the Philippines than in the United States. When He is your portion, you're blessed.

> You are my portion, O Lord. (Psalm 119:57)

> You will show me the path of life; In Your presence is fullness of joy; At your right hand are pleasures forevermore. (Psalm 16:11)

I was there to teach the people a better way—the Lord's way.

Another Mistake

Tom would be eighteen years old and have graduated from high school when I left for the Philippines. I wanted him to be with me, but I wanted him to know the limitations he would be under in a foreign country.

He asked if he could have a car and motorcycle, so I explained the situation to him. We would be there on tourist visas, and he would not be allowed to work. Even if he sold his pickup truck and bought some transportation there, he wouldn't be able to drive much, since the price of gasoline was twice to three times higher than in the states.

I also told him the money that was sent to us would have to go for the ministry, because those who sent us money wanted it to be used for that purpose. I made a mistake there. Money that is given to us to minister the Gospel should also be used to minister to our families. I wanted Tom and others to see and know the Lord the way He really is, but I was painting a different picture of Him altogether. That "poverty mentality" that others had sown into me was taking root.

I explained to him that the food would be different, the culture would be different, and there would be little for him to do with his time. Since there was nothing in my heart from the Lord that he would be working with me full time in the ministry, I thought he should be aware of those things.

I explained to him that I really wanted him to be there, but he should know the situation and make his own decision. He decided to go to Pennsylvania and live with relatives.

The Lord Intervenes

Before I left for the Philippines, the enemy was hard at work trying to create problems in our family. When Tom left for Pennsylvania, there was a serious breach between us. I went on with what the Lord had called me to do and trusted Tom to Him. As I was leaving this country, I called him and said goodbye. Shortly after I left for the Philippines, Tom left for the Air Force.

For a long time, I did not hear from him; then, about a year and a half later, I received a letter from him that was full of love. He wrote, "Dad, everything you ever said to me was right." I wrote back to him and explained that everything that I had said to him was God's Word, and because God's Word is right, I was right.

Because he was unhappy with me, when the Air Force asked him where he wanted to go overseas, he put down the Philippines as his last choice. However, that is where they sent him! The timing was perfect for both of us—for him, because by the time they sent him there, he was an E3; with that rank, they sent his car with him! As for me, I was now settled into my ministry and life there.

By that time, I was traveling frequently, and we were able to go to a lot of places and do a lot of things together. He was with me for more than a year.

At Clark Air Force Base, Tom was in a familiar culture, and since he was in the Air Force, he was working on his career and education. He had his car and a motorcycle, which he bought after arriving at Clark. These were important to him. He was also able to eat the food he was accustomed to. But the best part of it all was that he was with me. As far as we know, he was the only service person in the Philippines who could go home on his days off!

When Tom first got there, his roommate was involved in a lot of things, off base. I prayed about it, and in two weeks, Tom was given a new roommate, who was a Christian.

It was a blessing to go into Tom's room on base and see his roommate's Bible on his nightstand. It was well-worn, so I knew he read it. All things are possible with God.

> Ah, Lord God! Behold, You have made the heavens and the earth by Your great power and outstretched arm. There is nothing too hard for You. (Jeremiah 32:17)

Ready to Depart

Tom left for Pennsylvania in May, and two months later, in July, I began preparing to leave for the Philippines. I sold or gave away all that was left of our household possessions. When I contacted the Philippine Consulate in New Orleans to obtain a visa, they told me to send them the fee and my passport, and they would issue my visa.

I had been ordained, had incorporated the ministry, Faith By Love Fellowship, Inc., and had set up a corporate bank account. With most of my worldly possessions packed in two suitcases, I was now ready to go.

I had $800 in cash, no credit cards, and a little money in my bank account. My plan was to drive to Dallas to say goodbye to friends and family, to Idaho to say goodbye to family, to Washington state to visit relatives, and then to San Francisco to visit an aunt.

From San Francisco, I would drive to Phoenix to visit family and end up in Los Angeles, where I would sell the car, pay off the balance I owed on it, and use the rest for my trip to the Philippines.

I thought that I would reach Los Angeles in a few weeks, but my trip took almost three months! What wonderful visits! It was a great time. It was good to spend that time visiting everyone, because an aunt and an uncle died during my first trip to the Philippines. Because the trip lasted so much longer than I thought, I ran out of money, but the Lord had seen ahead and provided.

On the Sunday I spent in Idaho, I decided to go to a local church. It was a small congregation, and the pastor asked me to share. When they took the offering, I gave all the cash I had. However, I still had a few checks at the home where I was staying and some money in the bank; so I was not completely broke.

After I said goodbye and was walking to my car, the pastor called me back and gave me a check for $100. I didn't need the money then, but I would later.

A Different Kind of Offering

The next time I had an opportunity to minister, the Lord also gave me an offering, but in a different way.

I was in Spokane, Washington, at an uncle's house and was scheduled to leave for Seattle that day. I wanted to leave about 7:00

a.m., but wasn't able to because of all of the goodbyes and last-minute visiting. I was finally able to depart at about ten o'clock and was really irritated about leaving so late.

About 50 miles west of Spokane, I saw a car parked along the side of the road with its hood up. I noticed that the people were elderly, and as I passed them, the Lord put it into my heart to stop.

I tried to help the man get his car started, but couldn't. Then, he asked if I would take him and his wife back to Spokane. He said that he would have his son pick up the car.

Because I was so anxious to get to Seattle, I didn't really want to take them back, but of course, I did. This meant that I would lose another two hours, but what else could I do?

As we were driving, I told them I was on my way to the Philippines as a missionary. I began to share the Gospel with them, and they told me they were already born-again, so we had a nice visit on the way back to Spokane.

When we got to Spokane, it was noon, and the man asked if they could buy my lunch. Realizing that they were hungry and needed something to eat, I consented. The man was well-known and well-liked in the restaurant.

As we got up to leave, he took out two $20 bills, handed them to me, and said, "This is for bringing us back to Spokane." I handed the money back to him and told him I appreciated it, but I didn't want anything for bringing them back. We handed the money back and forth a few times until he finally took the money and put it in my shirt pocket. I then put the money in his shirt pocket.

Finally, he took the money, threw it on the table, and said, "Give it to the waitress." Then, he turned around and walked out. Since I knew he wanted me to have the money, I picked it up, put it in my pocket, and followed him to the car. Here again was money I would need later.

The couple's home was a small frame house. We parked in front, and when I looked at the man, I saw tears streaming down his face. I was surprised, and as I looked at him, his wife asked, "Do you want to know why he is crying?"

She said, "We were wheat farmers. When we retired, we sold our farm, moved to town, and bought this house. We were going to live on the money from the sale of the farm, but a Christian evangelist took our money. The reason he is crying is because today, he met a man of God who doesn't want his money."

The reason for all the delays was so the Lord could use me to restore that man. I didn't get to Seattle until late that day, but it didn't matter. The Lord wanted to restore him, and He gave me the opportunity to be the one He used.

Unexpected Provision

After almost two months, I arrived at my aunt's in San Francisco and the checkbook that I had my brother mail there for me. I had less than five dollars left, but the Lord had provided all that I needed. I never ran out of money, ate well, always stayed in nice motels, entertained, and had opportunities to bless others. From the time I left Tulsa until I got on the plane at Los Angeles, the Lord provided all that was needed ahead of time.

Four times on the trip to Los Angeles, I received finances from unexpected sources. Two I have just described. The third was from my home church. After I left Tulsa, they sent me $500. A few months after I arrived in the Philippines, they began sending me money each month and continued to do so for the next eight years. I do not recall the fourth provision.

According to what the Lord had instructed, I had not spoken to anyone about support. But as I went, all that was needed was always there, and I've had many opportunities and the provision to help others.

Soon after I left Tulsa, the Philippine Consulate in New Orleans mailed my passport and money to my brother's home with a letter saying that they would not issue a visa to me. Since I was on my way, I had a decision to make: Would I turn around and go back, or would I continue? Since I knew I had heard from God, I continued.

Every time something has come against me to stop me, the Lord has provided what I needed and given me the victory.

When I arrived in Los Angeles, I went to the Philippine Consulate and applied for a visa. They said that they would only give me a visa for fifty-nine days. When I asked them for a longer-term visa, they said no and this was all they would grant me. When I asked if I could extend my visa in the Philippines, they again said no; I would have to leave the country when it expired. I took what they offered and trusted the Lord to provide whatever else I needed.

The next thing I needed to do was sell the car. It seemed to me that I should wait before trying to sell it, but I was eager to get going, so I advertised the car in the paper. I had no peace, but I did it anyway. Not only was the advertising expensive, but also it was a waste of time. One man responded, but he didn't buy. Now that I had tried advertising and it didn't work, I didn't know what to do.

Then, a friend of a family member heard I had a car to sell. She came and looked at it, liked it, and bought it.

If I had just trusted the Lord and waited, I would have saved myself money and anxiety.

After I sold the car and paid off the balance, I had $1,800 left. Then, I remembered two debts I still owed to people in Phoenix, so I sent them their money. Both of these debts were more than fifteen years old, but I wanted to leave the country being right with everyone.

By the time I left Los Angeles, I had about $400 left in the checking account and $180 in cash, $50 of which was tithe money. As soon as I arrived in the Philippines, the Lord told me to give Mrs. Cruz the $50 tithe and another $120. That left me with about $400.

Within a few hours after I arrived in the Philippines, the Cruz family took me to a Bible study, and my ministry was launched.

Before I left Tulsa, I met a Filipino pastor at a meeting. As we talked, he asked me what I did in the ministry, and I told him I could do anything. He said, "No, no. I mean are you a teacher? Do you do crusades? What do you do in the ministry?" Well, I didn't know what God had called me to do, but whatever it was, I could do it.

Then the Lord spoke to Moses, saying: "See, I have called by name Bezalel the son of Uri, the son of Hur, of the tribe of Judah. And I have filled him with the Spirit of God, in wisdom, in understanding, in knowledge, and in all manner of workmanship, to design artistic works, to work in gold, in silver, in bronze, in cutting jewels for setting, in carving wood, and to work in all manner of workmanship." (Exodus 31:1–5)

We have the same Holy Spirit today and the same abilities. And am I ever glad to have Him. When I arrived in the Philippines and began my ministry, I needed all the faith and confidence I had expressed to that pastor.

CHAPTER 3

Important New Friends

The only public ministry I had ever done was giving a short message and an invitation at Christian movies. Once, my pastor said that since I was going to be in the ministry, I should preach in the church, so he invited me to give the message one Sunday morning. I don't know how good it was, but I ran half an hour over, and he never asked me to speak again! I then decided to hold a Bible study in our home. After the first night, I felt I had done such a poor job, I never tried it again. I invited our pastor to teach the Bible study from then on.

When the Cruz family took me to that first Bible study, I was asked to teach. I was doing so poorly, I remember looking down and seeing a crack between the boards and wishing it would open up so that I could fall through.

When I first started my own ministry in the Philippines, a group of ladies asked if they could sit in on a class I was teaching at the school I had opened. When they did, they were obviously disappointed and never came back.

But thank God, He is faithful, and has brought me through. A man once said to me, "You are a very good teacher." My response to him was, "It is all God. If it weren't for Him, all I would be is a pile of dust." A few days later the Lord said to me in a humorous manner, "Tom, if it weren't for Me, you wouldn't even be the dust."

Even though I was not qualified by human standards for the ministry, the Lord believed that I was. Just before my departure for the Philippines, the Lord told me what He wanted me to do when I arrived. He said, "When you arrive in the Philippines, I want you to start a school to prepare people for ministry."

The third day after I arrived, Mrs. Cruz took me to an area church rally and evangelistic `crusade in Dagupan City, which is about three and a half hours north of Manila.

The Lord had told me that He didn't want the school in Manila; He wanted to train the ministers in the environment in which they would minister, out in the provincial areas. He did not want them to become accustomed to the glamour and modem conveniences of a big city and then not be satisfied to live and minister in less affluent and comfortable surroundings.

The afternoon we arrived, I met a pastor from San Carlos City. His name was Cruz also, and as we talked, I found that my vision to start a school of ministry was similar to a desire he had. He wanted to set up a school to train Christian workers for the churches in his presbytery. Since everyone was busy with the rally, we didn't have an opportunity to discuss setting up a school then. However, I told him I would come back in a few weeks and talk to him about starting a school, which I did.

Chewie Enters My Life

While I was in Dagupan City, something important happened at the Cruz home in Manila. Their dog had puppies. The reason this was so important to me was because I was given one. Before preparing to leave for the Philippines, I had had three cats and three dogs. I enjoy pets, and since I would be living in an isolated area, a puppy would be a special blessing.

I was able to spend time with him, and we became attached. I named him Chewie, because he always wanted to chew on my fingers.

A woman offered to buy him, so they sold him to her. I missed him, but he was gone, so that was that.

The day before I left for San Carlos City to start the school was the Sunday before Christmas, a very busy day. I had been invited to attend one church in the morning for their Christmas program and the distribution of Christmas packages to the children in their area and then to spend the rest of the day at another church.

By the time I got home that night, it was late. When I went to my room, there was a sign on the door which read, "Be careful, puppy in the room." As I opened the door, there came Chewie bouncing across the floor and back into my life!

I was told later that when the woman got Chewie, she took him to a veterinarian for his shots. They also washed and trimmed him. But when she got him home, he was so unhappy she supposed he missed me, so she brought him back. Of course, I believe the Lord was preparing him just for me.

My Prison Ministry

While in Manila, I was invited to go to a prison to teach. There is a Bible school there, and all the students are prisoners who have been sentenced to death and are awaiting execution. They asked me to minister, and it was great to see men who had once been hardened criminals but are now loving and gentle, hungering for God and His Word.

At the end of the teaching time, they brought a man to me. They said his name was Mario, and he was scheduled for execution that afternoon at two o'clock. (It was already about 11:30 a.m.) His head and a spot on one of his legs had already been shaved to receive the electrodes. They said that he wanted me to pray for him.

As I started to pray, I asked the Lord to strengthen him, encourage him, and help him go through the execution. When they heard that, they said, "No, no. He doesn't want you to pray that! He wants you to pray that his sentence will be commuted to a life sentence so he will be sent to another prison. There, he will be able to preach the Gospel."

I learned something that day. I learned to find out what the person you're praying for wants God to do before you start to pray.

> So Jesus stood still and commanded him to be brought to Him. And when he had come near, He asked him, saying, "What do you want Me to do for you?" He said, "Lord, that I may receive my sight." (Luke 18:40–41)

Here was a blind man, who obviously needed his sight restored, but Jesus did not assume anything nor did He speak for the man. He had the man state specifically what he wanted Him to do. And then, he was granted exactly what he wanted and had the faith to receive.

> Then Jesus said to him, "Receive your sight; your faith has saved you [made you well]." (Luke 18:42)

When they told me that, I prayed again and asked the Lord to commute his sentence so he would live and declare God's power and goodness in another prison.

About a month later, I was invited to go to that prison again and minister. I looked for Mario but didn't see him. I became anxious to know what had happened to him. When the teaching was over, some of the men came up and began talking to me, but what was really on my mind was Mario. I didn't want to ask them where he was. Hadn't we agreed in prayer that his sentence would be commuted and he would be sent to another prison?

> Therefore I say to you, whatever things you ask when you pray, believe that you receive them, and you will have them." (Mark 11:24)

So I stayed quiet and waited.

Finally, one of the men said, "Did you hear what happened to Mario?" I said I hadn't. He said, "His sentence was commuted, and he has been sent to another prison."

My Finest Meal

One night, the Cruz family asked if I wanted to go to McDonald's for dinner. What a blessing! I was really hungry for American food. As we were eating, I noticed some little children standing outside, looking in. They looked hungry, so I bought each of them a hamburger.

Apparently, they had never had one before, because when I gave them the hamburgers, they sat down on the sidewalk, opened them, and started going through them. I'll never forget it.

When they got to the pickles, they picked them out, chatted together for a minute, and then threw them away. They didn't know what pickles were, and they were not going to eat them. Finally, after inspecting their hamburgers thoroughly, they ate them.

What a wonderful opportunity to share with those children!

> Command those who are rich in this present age not to be haughty, nor to trust in uncertain riches but in the living God, who gives us richly all things to enjoy. (1 Timothy 6:17)

God wants us to enjoy our lives.

I have eaten some wonderful meals in beautiful and expensive restaurants, but I never enjoyed a meal more than I did that night in McDonald's as we sat eating our food and watching those little children eat theirs. It was rich, and we enjoyed it.

A Lesson from Two Little Boys

This reminds me of a time, years later, when I was distributing tracts in the University District in Manila. Not only universities are there, but also elementary and high schools. It was early in the morning, and I was distributing tracts to the students on their way to school.

As I stood there, a little boy, about six or seven years old, walked by. He was filthy and dressed in rags, obviously living in an alley somewhere. (How sad to see children or anyone else in that condition.)

After a while, another little boy, about the same age, walked by. What a difference! The second boy was spotless. He obviously ate well and had good coloring. His hair was damp and freshly combed. He had on a white shirt and dark blue knee-length shorts with suspenders. He wore white knee socks with little black leather shoes. All his clothing was freshly laundered and pressed. And he was pulling a little book carrier with wheels. He was really cute.

After the second boy walked by, the Lord spoke to me and said, "Do you know the difference between those two little boys?" Then, He said, "The father they have."

How can we ever doubt our heavenly Father's love, and His desire to provide for us?

> The destruction of the poor is their poverty. (Proverbs 10:15)

> The thief does not come except to steal, and to kill, and to destroy. I have come that they may have life, and that they may have it more abundantly. (John 10:10)

> The blessing of the Lord makes one rich, and He adds no sorrow with it. (Proverbs 10:22)

Poverty and destruction come from the devil, and abundant life comes from God.

> And in that day you will ask Me nothing. Most assuredly, I say to you, whatever you ask the Father in My Name He will give you.

> For the Father Himself loves you, because you have loved Me, and have believed that I came forth from God. (John 16:23,27)

Doors Open

As promised, I went to San Carlos City and spent a few days with Pastor Cruz and his family. After returning to Manila, I prayed, and even though I did not get a discernible response from the Lord, I decided to go there and start the school.

Meanwhile, Mrs. Cruz, whose family I had stayed with in Manila, asked me to take over her hospital ministry. In her hospital ministry, she traveled to hospitals throughout the Philippines and ministered to groups of doctors and nurses. It would have been a wonderful vocation, ministering to those who are ministering to multitudes of people at very critical times in their lives, and all of the travel would have been great.

On the other hand, San Carlos City was about four hours from Manila and remote. There would be no television or radio, few modern appliances and conveniences, and no recreation or entertainment. It would be an austere life with none of the things we Americans are used to. But I knew that the hospital ministry was not what the Lord had called me to do. He had said to start a school.

I had arrived in Manila on October 20, and I was scheduled to leave for San Carlos City about December 20. Since my visa would expire just before I was to leave for San Carlos City, I had to do something about it before I left. I thought I should go to the Immigration Department to see if I could get an extension. They had told me in Los Angeles that I would not be able to extend in the Philippines, but what had always been in my heart was that I was to go to the Philippines long term. I decided to go to Immigration and see what I could do.

When I mentioned this to a few people, they told me that if I got a visa extension, it would take about three days of standing in line before everything would be completed.

Nothing Is Impossible

The day before I planned to go to Immigration, a woman at a Bible study asked me what I was going to do the next day. When I told her I was going to the Immigration Department to see if I could get my visa extended, she said that she knew a woman who worked there. She said that she would call her, and I should be sure to see her.

The next afternoon, I went to the Immigration Department. It was a big building full of people. The offices and corridors were jammed. I asked for Mrs. Etta Veneration and was directed to her office. She was expecting me and invited me into her comfortable, air-conditioned office.

After I explained what I wanted, she asked for my passport and called a woman named May into her office and asked her to take care of this for me. She then sent out for refreshments, and they brought me orange pop and cookies. As we continued to visit, one of her sons came in. I was able to share the Gospel with him, and he prayed with me to be born again.

After this, May came back and asked me to go with her to have my photo and fingerprints taken. Then, she took me back to Etta's office.

After a short wait, she returned with my visa, and I was ready to go home. The total time there was about an hour and a half!

Before I left, May asked for prayer for her two children who were ill. Later, I found out the Lord had completely healed them. The Lord provided what I needed and blessed the two women who helped.

During the next three years, whenever my visa expired, all I had to do was to take or send my passport to Etta, and she got it extended.

A Visit to Hell

I will never forget the time during that first three years when I tried to get my visa extended myself. That morning, I arrived at

Immigration before Etta got to work. Because the offices were open, I thought I would handle the visa myself. What a nightmare!

They could not find any of my records and said I would have to refile all my paperwork. That meant filling out all new forms and getting new photographs and fingerprints. Not only would I have to do all that, but also the people I was talking to treated me like a criminal. I felt like I had stepped into hell and demons had been unleashed against me.

I got out of there and went to Etta's office. She was there, and she had my visa ready in no time with no problem at all. That is when I knew for sure that the Lord had provided Etta to help me, and He wanted me to receive her ministry, which I was glad to do. I didn't want to go through that again.

The next job was gathering up the things I would need when I got to San Carlos City—towels, washcloths, and a 110-volt transformer for my hair dryer—but I didn't have the money. One day after a Bible study, a group took me to lunch. When we got back to the Cruz home, three of the attendees asked if they could come in for more ministry. I said yes, and when they left, they gave me an offering of 1,400 pesos, which at that time was about $175. I now had the money I needed.

Life in San Carlos City

The next thing I had to do was get to San Carlos City. I had a lot of things to take, and unknown to me at the time, I would also have a new puppy to take along that would have to take a few walks before we got there. There was a bus I could take to within twenty miles of San Carlos City, and then, I could take a jeepney the rest of the way.

It would be a difficult trip by bus. As I was making my plans, it just so happened a man I had met at church decided to visit his mother in a place beyond San Carlos City. He didn't have a car, so his cousin agreed to take him. Since he knew about my trip, they asked if I wanted to ride with them. Blessing upon blessing!

When I arrived in San Carlos City, Pastor Cruz asked if I would like to sleep in the balcony of the church and eat in the house with them. When I agreed, they showed me the bed that I would be sleeping on. It was a solid wood bed with no mattress.

When I saw that, I asked if there was any place where I could buy a mattress, and they said that I could get one in Dagupan City. Pastor Cruz drove me there, and we picked one up in his Volkswagen van. The money those people had given me was all I needed for the move.

Before I left the United States, the Lord showed me how to handle the transfer of money to the Philippines. He directed me to open a bank account in Tulsa and then to tell anyone who wanted to send money to send it to the Tulsa ministry address, which was my brother's home. Once a month, my brother was to send a telegram and tell me how much money had arrived, and I would write a check over there.

At that time, it could take up to thirty days to transfer money from an account in the United States to an account in the Philippines, and it was expensive. If you write an American check in the Philippines to be sent to the United States for collection, it would take sixty to ninety days to get the money.

When I first got to the Philippines, I cashed my checks through Mrs. Cruz. She said that they knew her at her bank and would cash my checks right away; but after I went to San Carlos City, I wouldn't be able to cash them through her any longer. However, when I arrived, the Lord's provision was ready. A wealthy man who attended Pastor Cruz's church asked if he could cash my checks. He said that he would give me the money right away or at the longest in two or three days, and he would wait the two or three months for the transfer. Besides that, he took care of whatever transfer expenses there might have been. What a blessing!

Some people in the Philippines have dollar accounts in their banks. These accounts are a good hedge against inflation, and when they travel abroad, they have dollars to spend.

Only once in the history of this ministry have I issued a check that was returned for insufficient funds. And I know the money was in the bank to cover the check. When the check was returned, the man gave it back to me, and I gave him another check. When I looked at the back of the returned check, I noticed that it had a Hong Kong stamp on it and that bothered me. I knew that a lot of American money was being blackmarketed to Hong Kong. By God's grace, I have tried to deal in every situation in absolute righteousness—no bribes and no black market.

> Therefore submit yourselves to every ordinance of man for the Lord's sake. (1 Peter 2:13)

I don't think the man who was cashing my checks was black-marketing the money, but possibly someone in his bank was. Anyway, I was disturbed about it, and a short time later, a woman who lived in Manila asked if she could cash my checks. Not only that, since I was ministering to her, she said that she wanted to give part of her tithe to me. She did this by giving me two extra pesos for every dollar of the checks that she cashed for me. When she did that, I was still not getting as much as those who were selling on the black-market, but when the black-market stopped operating, I was still getting the extra money and the others weren't.

> Lord, who may abide in your tabernacle? Who may dwell in Your holy hill?
> He who walks uprightly, And works righteousness. And speaks the truth in his heart. (Psalm 15:1–2)

She also gave me my money right away and paid the transfer expenses. That lady continued to cash my checks for another eight years although in the later years, she did not give the extra money.

I would like to share with you a way that the Lord really blessed her. She was the manager of a large restaurant that specialized in fried chicken. As we were having a Bible study one day, she mentioned that business at the restaurant was declining, and she was very concerned. Since she was a Christian and in covenant with God, I knew

He would bless everything she put her hand to. I told her that even if business was slow in Manila, God could bring everyone in the city who wanted a chicken dinner to her restaurant. I told her to look to God, believe His Word, and He would provide.

A couple of months later, the Quezon City Government contacted her and gave her an order for over 22,000 complete chicken dinners. Jesus said in Matthew 21:22, "And whatever things you ask in prayer, believing, you will receive." In Deuteronomy 15:4, the Bible says, "Except when there may be no poor among you, for the Lord will greatly bless you in the land which the Lord your God is giving you to possess as an inheritance." It's God's will that there be no poor in His Kingdom. "The blessing of the Lord makes one rich, and He adds no sorrow with it" (Proverbs 10:22).

Twenty-two thousand chicken dinners make up for a lot of slow business days.

When I incorporated the ministry in the Philippines, got the Bureau of Internal Revenue tax exemption, and got my 9-G (long-term) visa, because I took a strong stand for righteousness, the Lord was able to bless me.

In 1987, the Lord told me it was time to get these things done. First, He gave me favor with an outstanding law firm who would not take any new clients, but they took me. When I first talked to the attorney who was to handle the account, I told him everything had to be done in a completely legal manner. There were to be no bribes or anything else illegal.

He replied, "But you don't know how long it could take," meaning that without these things, it could take a very long time trying to get everything done. I told him that I didn't care. Everything had to be right.

When it was all done, I had everything I wanted and needed, with an extra bonus. The day that my 9-G visa came through, the law firm was ecstatic. I had been given a five-year visa. They had never heard of that before. Usually, they are issued for one year. I'm not saying no one else has ever gotten a five-year, 9-G visa before, but they had never heard of it.

Another thing that happened was the billing from the law firm. When they billed, they would always give me a statement that listed the work that had been done and the charges and then at the bottom was the total. On every statement, they gave me a 50 % deduction, always charging me half of the cost. I don't know why they did this. When I asked them, I never got an answer. The cost of getting all that legal work done was very expensive; so it was a large savings!

When the Lord told me to do these things, He at the same time gave me 5,000 pesos to give them to begin the work. And every time I was given a statement, the Lord provided the money to pay it. Working for God is great.

We are the light of the world and the salt of the earth and the Lord expects us to act like it. Everyone who comes in contact with us should be coming in contact with Jesus.

The Foreigner in the Balcony

As I said previously, when I got to San Carlos City, I slept in the balcony of the church. In the Philippines, many churches have early morning worship services the week before Christmas. Because my first week in the church balcony was the week before Christmas, I had to get up early. Of course, I wanted to get up and be ready for the service, but I had to really get up early because some of the children would come ahead of time so they could look up in the balcony to see the foreigner! The first morning they came in, I didn't expect anyone that early, and I was not dressed properly to receive guests. Believe me, I got up much earlier from then on.

In November, some friends from Texas sent me $250. With what I already had, I was all right that month. The month of December was when I received the 1,400-peso offering for San Carlos City.

Even though the Lord was providing, for a while, I had to battle with lying thoughts. Since I had not talked to anyone about support, I didn't know where it would be coming from. When I started the school, I could not charge the students anything for tuition or expenses, because that would be violating my visa. I would have to be the one who financed the school.

And even though I was eating my meals with Pastor Cruz and his family, I believed it was necessary to give them money for the food I ate and give Mrs. Cruz something for the extra work I created for her. One of the many reasons that I love the Lord, and enjoy serving Him, is because He is so gracious, and He provides what we need so that we can be too.

Questioning Thoughts

In December, when I had just enough to live on, questioning thoughts began to run through my mind. I began to think about my family and friends. *They know I'm over here. Don't they know I need help? They do know that I need help.* Then, I started thinking; *They don't care about me. They've forgotten me.* I started thinking that I had been deserted. Fear began to move in, thoughts that I was going to fail and perish. Then, feelings of despair began to flood in.

I was opening the door wider and wider to the enemy by receiving his thoughts and not standing fast on God's Word. I began to realize I had to get my thoughts under control, so I began meditating on God's promises, and the Lord pulled me out of that pit.

The Lord Jesus said in His Word,

> Therefore do not worry, saying, "What shall we eat?" or "What shall we drink?" or "What shall we wear?"
> For after all these things the Gentiles seek. For your Heavenly Father knows that you need all these things. (Matthew 6:31–32)

Through December, the Lord provided everything I needed, and by January, money began coming in regularly from the states.

I have always offered money to the people I have stayed with to cover my expense to them. Some needed the money and received it, while others declined. God has not sent His ministers into the world to be a burden to others; He has sent us to help others with their burdens.

Preparing for School

There I was, ready to start school, and I had no idea what to do. I decided I needed some time to prepare, so I scheduled school to begin on January 18, 1982.

I had a good foundation in God's Word, but I didn't have any idea what a school of ministry should be. Not knowing what to do, I sat down and made a list of the classes I'd taken in Bible school. From that list, I set up the curriculum and prepared my lesson plans, prayerfully. I spent the next three weeks preparing for school.

I didn't give any thought to students. In fact, I made no effort to draw students to the school. I was so involved in getting ready for class that I never thought about students! But one at a time, they began to come and ask if they could attend. Pastor Cruz told many people about the school, so I know some of them heard about it that way.

I'll never forget the evening I met Digna, the first student to arrive at school.

She was eighteen years old, but she looked younger, being about five feet tall. She only weighed about eighty pounds. Digna had come to school with fifteen to twenty pounds of rice in a cloth and her clothes in a small travel bag. She was timid and would hardly speak, but what a blessing and mighty woman of God she was and still is!

All the students except one in that first class lived at school. On weekends, the other students would go home, but Digna couldn't. She didn't have the money, and her home was about three hours away by bus, so she stayed at school on the weekends and spent most of that time alone. I know how lonely she must have been, but she held fast.

Her father once was badly gored by a bull and was in the provincial hospital in Dagupan, about forty-five minutes away. She spent each evening with him for weeks, but she never missed a class. She never missed her daily three-hour ministry assignment, and she never missed a chore.

Our Success Rate

After Digna left school, she went home and went to work in her home church, and the last I heard, the three male graduates from the first class are still in full-time ministry sixteen years later. One is an evangelist and two are pastors. Both of the pastors are anointed evangelists. I don't know how many students from the second and third classes are in full-time ministry today.

One of the reasons for the success rate is that the Lord prepared and brought the students He wanted to train. Each of the three classes had six students. Now, I can see why the classes were small. When you are training ministers, I believe that you need to spend a lot of time with each person. You are ministering to individuals, not to a group. The students take this concept into their own ministry, because that's what they've been taught. That way, ministry becomes an individual relationship between the shepherd and each sheep. Jesus always had time for each person who came to Him. And He also trained the apostles in a small group.

The second class was one year in length, and Digna came back for the second half to train the students in evangelism and outreach. She brought her sister Fe with her. What a wonderful anointing Fe has for ministry to children! What a great honor and privilege it is to be called by the Lord to teach and train these people.

Practical Experience

Other reasons for the success of the students were the Lord's direction to train them to do the work of the ministry and to give them someone to be accountable to while they were in training. The three students who graduated with Digna, Daniel (pastor), Juhn (pastor), and Bogs (evangelist) went straight into their own ministries after school.

The Lord revealed that I had to give the students practical ministerial experience in school if they were to be successful ministers. An American denomination that has many three-year Bible schools in the Philippines made a survey and found that 85 percent of those who had graduated from their Bible schools were not in any type of ministry afterward!

One Saturday afternoon, some of our students took students from a three-year Bible school door-to-door witnessing with them. Some of those students were close to graduation. Later, some of them told our students they had learned more that afternoon than they had in all their years in school. I don't think that is true, but I think what impressed them was working with the Holy Spirit and putting what they had learned into practice.

The first three months, the students did not go out to minister. We held classes in the morning, and the afternoons were used for Bible reading, study, and prayer.

One of the reasons Jesus was a perfect minister was because He was full of all the Word of God. All the Word of God was abiding in Him. All the Holy Spirit had to do was reach into His heart and bring forth the right word at the right time to minister to what every person or situation needed.

Salvation Classes

In my Bible reading, I ultimately came up with more than 300 scriptures that pertain directly to salvation. Of course, all scripture pertains to salvation, but these are scriptures that speak directly to man's need for salvation and what God has provided through Jesus Christ who supplies the answers to that need.

During salvation class, we studied all our scriptures on salvation. We filled our hearts with those scriptures.

> And they went out and preached everywhere, the Lord working with and confirming the word through the accompanying signs. Amen. (Mark 16:20)

The word "them" in this verse in the King James Version is not in the original text; it was added by the translators. The Lord works with and confirms His Word.

During the next class of students, some people from a local church asked if I would hold a salvation class for them. Several of them were college students, so this class was taught during their summer break.

For this group and our regular class, I made up a game called "Gospel." It was the same as bingo, except the cards were seven squares wide and seven squares high. I put all our scriptures, book, chapter, and verse on little cardboard cards, which I put in a bowl. In each space of the "Gospel" cards, I wrote a scripture reference, book, chapter, and verse, with each card being different.

I gave each student a "Gospel" card. Then, I pulled a card out of the bowl, found that scripture in the Bible, and read it to them without telling them what it was.

The object was for the players to hear the scripture as I read it and then for them to find the right square on the "Gospel" card and mark it. The first one to have a full line won.

The reason I did that was so our hearts would be full of God's Word. Then, as we ministered, the Holy Spirit could reach into our hearts and minister to each person exactly what He wanted to. We want to be able to serve all the Lord has provided on His banqueting table.

> So, as much as is in me, I am ready to preach the gospel to you who are in Rome also. (Romans 1:15)

Another reason was that when the Holy Spirit would bring a scripture to our remembrance, we would be able to find it in the Bible.

> But the Helper, the Holy Spirit, whom the Father will send in My name, He will teach you all things, and bring to your remembrance all things that I said to you. (John 14:26)

One thing I taught the students was to never quote scripture from memory when preaching, teaching, or witnessing. I taught them to always find the scriptures in their Bibles and read them to the people and, when witnessing, to always show the scriptures to the people and let them follow along while the scripture was read. The Bible was always our textbook.

Witnessing Practice

At the end of the second month, we began to practice witnessing each afternoon for about two weeks. During this time, the students would witness to me and to one another.

We would set two chairs in front of the classroom, and one of us would sit down in one chair and a student would walk up with their Bible and ask if they could talk to us about Jesus. We would say yes, and the student would sit down and lead us to the Lord and pray with us to receive Jesus. Each student would take a turn.

After a few days, we would begin to resist by trying to lead them off on "rabbit trails," asking irrelevant questions; trying to tell about "our" religions, what we believe, or what other people believe; or just trying to get them off the subject. I taught them not to let themselves get tangled up in such conversations with people. They were there to lead them to Jesus and not to let anything interfere with that. As an example, someone might ask, "What about the people out on the islands or up in the mountains?" I taught them to respond, "Sir or ma'am, God will take care of them. I'm here to talk to *you* about Jesus."

After that, we spent two weeks practicing preaching. Each day, each student would get up and give a 15- to 20-minute evangelistic message and then pray with everyone to be born again. I taught them to always have a second or third witness for every scripture they used. Any time they were ministering and someone wanted another scripture to confirm what they were saying, they should always be ready.

Also, toward the end of the first class of students, I had them prepare Bible studies and teach them to the class.

At the time, it was a lot of hard work, but looking back, it was also a lot of fun!

Preaching For Real

The first day of outside ministry was the day the students were to preach in the streets. This was the only scheduled day of street preaching during school.

We went in a group. I permitted the students to select the place where they wanted to preach. During street preaching, we tried to have someone with us who played the guitar.

When we got to where the student wanted to preach, the one who played the guitar would play, and we would all sing a few Christian songs. As we sang, a crowd would gather. When I felt it was time, I would give the student the cue, and he or she would begin to preach. After he was finished preaching, he would pray with the people to receive Jesus. After that prayer, the student would tell the people that if anyone wanted prayer for anything else to let us know. Then, all of us would move through the crowd and pray for them.

When the first class went out, Digna was the first to preach. She chose a spot on a main street. When she started to preach, her lower lip was trembling. She was frightened, but she confidently continued and did a good job.

The second student to preach was Daniel. He led us to the big Catholic cathedral and up to the massive front doors. It is a big place, and people are constantly coming and going through those doors. He said that he wanted to preach there. I told him I didn't think they would like that, so he went a short way down the sidewalk and preached at the entrance to the park. Another preached in the park. I don't remember where the others preached.

One thing I saw right away was that many of the people were paying more attention to me than to the ministers. I realized then that I would not be able to go out on a daily basis with the students. Also, they needed to learn to depend on the Lord's presence, help, and guidance, not mine.

Settling In

I continued to sleep in the church balcony and eat my meals with the Cruz family, but it became obvious this was not a long-term solution to my housing needs. After a few weeks, Pastor and Mrs. Cruz told me that they had decided I should sleep in their bedroom, and they would sleep with their children. I didn't want to do that, but neither did I want to offend their hospitality.

Filipinos are a very gracious and hospitable people. Everyone I have ever met who has been in the Philippines would like to go back, and they all have this same thing to say about the people. I thanked Pastor and Mrs. Cruz for their thoughtfulness but declined their offer as firmly as I could; however, I was a guest in their home, and they insisted, so I finally agreed. I knew I had to make other living arrangements, but I was so concerned about being ready for school I didn't give it much thought. I also believed that God would provide for that need.

A few weeks after school started, a family who owned a large house up the street agreed to rent their upstairs to me.

I took a small room that had a table in it and made that the kitchen. (Because of the heat and the fact that most Filipinos cook with wood, their cooking is done outside.) I bought a gas hot plate and propane bottle to cook on.

I rented a refrigerator, but I couldn't get it up the stairs, so I put it in the church and carried the things needed for meals to the house.

CHAPTER 4

Living in a Hostile Environment

Until a few of weeks before I left for the Philippines, I didn't know where I would be, what I would do, or how I would live, except I thought I would live in poverty. Before I left Tulsa, my mother asked when I would be back. The Lord had not said anything to me about returning, so I said, "I don't know. Maybe I'll never be back." That was the wrong thing to say. It really upset her and proved how little I knew about the Lord.

When I moved to San Carlos City, as far as I knew, it might be the place where I would spend the rest of my life, but I was there for only eight months. I was having a great time! I was where God wanted me to be and doing what God wanted me to do. And all my needs were met, which was great, but it was a hostile environment. Murder was not uncommon.

A man and his family who lived down the street were sitting out in their front yard one evening. When he got up and went into the house to get a drink of water, there was a man in their house robbing it. When he saw the burglar, he ran. The next evening, the owner of the house was killed in his home, probably so he couldn't identify the robber. There were nights I could hear gunfire nearby. There were no telephones, few police, and no help if needed; but I believed the Lord would provide, and He did.

Before I got there, there was a time when the streets in San Carlos City were almost empty during the day, because there was so much violence and fear. By the time I got there, that was no longer the case.

There was a medical school in San Carlos City. One of the students was from Africa, and his wife and family were there with him. There were nights when the electricity would go off, and you never knew for how long. Once when it had gone off, this family heard a noise downstairs at their front door. When they looked out the window, they saw some men trying to get in.

The family became very frightened because they thought if the men got in, they would kill them. I agree, I think they would have. They started calling out for help, and the men kept on trying to get into the house.

Then, the family began crying out to the Lord. They said, "Lord Jesus, save us! Lord Jesus, save us!"

Suddenly, the lights flashed back on, and the men ran away.

> I cried out to God with my voice—To God
> with my voice; And He gave ear to me.
> In the day of my trouble I sought the Lord.
> (Psalm 77:1–2)

Special Blessings

The Lord often gave us special little blessings. One Sunday, as we walked home from church, someone said, "I would really like to have some chocolate ice cream." A few minutes later, the man who was cashing our checks and his family pulled up beside us and handed us a plastic bag out of their car window. Inside was a package of chocolate ice cream!

Another time was when I was in a ministry office in Manila about noon. I was getting hungry, and I remembered there was a Kentucky Fried Chicken restaurant at the end of the street. I was thinking how good it would be to go there and have some chicken, but I didn't have the extra money to eat out. A few minutes later, Steve, one of Mrs. Cruz's sons, came in and said, "Let's go to Kentucky Fried Chicken for lunch. I'm buying." Praise the Lord!

One day, I was walking down the street near Mrs. Cruz's house and saw four men walking abreast toward me. I didn't think anything about them. When I was about forty feet from them, another of Mrs. Cruz's sons pulled up in their pickup truck and asked if I wanted a ride. I got in, and we drove away.

Sometime later, I started to think about that situation. I believe the Lord showed it to me so I could see His protection. All of a sudden, I could see everything vividly. I could see how the men were positioned across the street. I could see their faces, and I could see their intent. They were going to attack me!

Not only did the Lord deliver me from their intent, but also He delivered me from the fear of their intent. Often in the Bible, when God's people were in a situation that was alarming or could alarm them, Jesus would say, "Do not fear," "Do not be afraid," or "Peace." He never wants us to be fearful or not in peace.

When I have been assaulted, I was unaware of what was happening except once. My skin has not been broken, nor has a mark been left.

One day when I was in Baguio City standing on the main street passing out tracts, a Caucasian American stopped to visit with me. He said his name was Wade Phillips and that he was a teacher at the military dependents' school at Clark Air Force Base. He was a nice guy, and I enjoyed our visit. After we talked for a while, he said goodbye and left, and I never saw him again. A short time later, the Lord said it was time to move up into the mountains to Baguio City. When I asked him when I should move, I believed that He said June 24, which was a Sunday. Everything was not ready so I couldn't move that day but wanted to go there and do what I could, so I loaded some tracts into my shoulder bag and took the bus to Baguio City. As I stood there passing out tracts and looking to my left, someone suddenly grabbed me from my right side and pushed me around. When I looked to see who it was, I saw it was a Caucasian man and thought it was Wade Phillips just kidding around; so I said, "Hey, Wade. How are you doing? What's happening?" Then, the man burst out in a foreign accent, "What are you trying to do, make rice Christians?" Then, I realized he was not Wade. The man turned and walked on up the street. The way that it happened, I was not alarmed or angry at all.

In the Philippines, I usually tried to pass out tracts wherever I went. When I got on a bus, I gave tracts to all of the passengers. On one trip to Manila, I had done this, and as I was getting off the bus, someone called to me. As I turned to respond, a man threw a wadded-up tract at me and hit me in the face. Again, the way that it happened, there was no alarm.

The next time was when I was again standing on the street in Baguio City passing out tracts. I had just given a tract to a man from India or that region and his wife and daughter. The Lord has placed a special love in my heart for those people and a desire to see the Gospel ministered to them. As they were walking away up the street, I watched them to see what they were going to do with their tract. Suddenly, I was struck with a tremendous blow on the right side of my head. As I turned, a Filipino man was moving away from me. He stopped about ten feet away and began screaming things at me in another language. He was holding a leather jacket in his hand, and I knew he had hit me with it. When he hit me, I could feel something heavy and hard. I think he probably had a gun in the jacket since I knew many men in that area carried weapons.

But again, the way that it happened, I was not alarmed; however, that time I was a little shaken after it was over. And later, from time to time, thoughts would come to me, "Why did this happen to me? I'm doing the Lord's work?" Then, I realized those thoughts were coming to me from the devil. He was trying to get me to take those thoughts and be offended at God.

Another time was when I had just walked through a barangay (a barangay is a small village or district) and was walking along a dirt road on my way to some schools. About a quarter of a mile from the barangay, I saw a group of about twelve men and a few women on the road, outside a building. As I got closer, I could see they were drinking, and some were drunk. As I walked through them, some of them tried to get me to take a drink, and some were jostling me a little. As this began to happen, an older man whispered to me, "These are barangay people." I knew that he was saying to me that these were local people, they were harmless; there were no NPA (New People's Army rebels) in the group. I got through them and went on to the schools. As I was walking back down that road and got back to that area, they were still there. They seemed to be drunker, and I could see they were meaner. As I got closer, I thought about how I was getting ready to run a gauntlet. As I started through them, I just looked straight ahead and kept walking. The atmosphere was now much more serious. No one touched me, but just as I got past them, I heard feet running behind me and then a scuffle. I did not look back so I don't know what happened. I think someone was running after me, and some of the other men grabbed and subdued him.

A while later, I was back in that area ministering at those same schools again. That morning as I was teaching, there was a helicopter so near that I was having a difficult time making myself heard. When I looked out the window, I saw that it was a military helicopter making low sweeps over some nearby underbrush as if they were looking for someone. I believe there were NPA in the area that day, and the Lord had sent His protection.

One time, something happened that was kind of funny after it was all over. The area where I lived in Baguio was very dangerous. Two taxicab drivers had even been killed on our street, and because of that, some taxi drivers would not bring Filipinos to the area after dark. When the house I lived in was built, the landlady had a high concrete wall built on the street side of the house with two or three feet of barbed wire on top of that. To enter, you had to go through heavy steel gates with sharpened steel bars over them. Around the rest of the house, she had a six-foot cyclone fence with barbed wire on top. It looked like a prison, but all that protection was comforting.

One Saturday night, about midnight, I was lying in bed sleeping when all of a sudden I was awakened by a lot of automatic gunfire and men screaming. I could tell it was right out in front of the house, just a short distance away. What a way to wake up! I didn't know what was going on. Were they breaking into the house, was it a gunfight, or what? I admit I was alarmed so I started to repeat the twenty-third Psalm in my heart, and I immediately fell asleep. The next morning when I woke up, I had forgotten all about it. Later that morning, I stopped by to see the landlady about something, and as we were talking, she mentioned the commotion the night before and then I remembered it. I asked her what it was, and she said that it was the Philippine constables who lived across the street. She said that every once in a while, they would get drunk and do that.

Our houses were built on the side of a mountain, so I'm sure that sound carried a long way through Baguio. What was good about it was that not only was I aware there were fully armed Philippine constables living across the street, but also a lot of other people were.

Another thing that happened that was kind of interesting was the night a man on a bus that I was on began to threaten the other passengers with a .45 automatic pistol. When he got to me, he pointed it right at me. As he was pointing the gun at me, the conductor apparently saw what he was doing and told the driver to stop the bus.

He then walked to the man and told him to get off the bus, which he promptly did. The Bible says in Proverbs 28:1b, "The righteous are bold as a lion." As bold as that conductor was, he may have been a Christian.

There were other things that happened, and I will tell you of some of them later in this account.

Preaching to the Rebellious

When I first got to San Carlos City, the Lord said, "You will preach the Gospel to those in rebellion to this government." When He said that, I was alarmed. I thought, *Oh, no. Somebody's going to be kidnapped.* I knew the NPA kidnapped and even killed people.

Over a period of time, the fear of kidnapping faded away, and I forgot about what the Lord had said. About three years later, in late 1984, the Lord told me to put the Gospel in the newspaper.

I chose a large newspaper, which happened to be very anti-Marcos at the time. It was popular and was distributed throughout the Philippines. Even a nonpolitical Filipino told me that she read that newspaper.

She said that the other national newspapers were censored and biased, so she wouldn't read them. That seemed to be the general consensus.

I asked Philip and Maribel, two of my Filipino friends, if they would find that newspaper's office and see if they would accept my tract for printing. The newspaper agreed, so we bought half of page three. When they asked me what day I wanted to print the tract, I chose a day that seemed good.

It is a morning newspaper, so I got up early the day our article came out and prayed in the Spirit until midmorning for the people who would read the tract.

After my prayer time, since I had not had an opportunity to check the layout and text before printing, I went downtown and bought a paper. I took it home and checked the text word for word. It was perfect!

As I closed the paper, I noticed a bordered box at the bottom of the front page. I didn't know it, but the next day was Ninoy Aquino's birthday (the slain senator who became a national hero). And that box listed the antigovernment activities that were scheduled for the next day to commemorate his assassination.

Then, the Lord said, "This is what I meant when I said you would preach the Gospel to those in rebellion to this government."

It had been the perfect day to put the Gospel in the newspaper! It was the day those in rebellion, whether actively or in sympathy, would be reading that newspaper to find out the activities for the next day. Even average Filipinos would be reading that issue.

Not only did we get to minister the Gospel through the newspaper, but also we got to minister to the people at the newspaper. They said that they have never done business with anyone like us, particularly in the promptness with which we paid our bill. That was another time I had opportunity to learn that God's infinite ways of working are not limited to my finite reasoning.

Then, there was the afternoon after I had moved to Baguio City that the Lord put into my heart to "Take tracts and go to the south." I put tracts into my shoulder bag, went to the bus station, and took a bus south toward Manila.

After we had traveled about an hour, we came to a road block that the Philippine constables had set up. They were stopping all of the buses going north and searching them.

I knew that the Lord wanted me to give tracts to the people on the buses they had stopped, so as they got on each bus and searched, I got on with them and gave everyone a tract. After about an hour and a half, the PCs took down the roadblock and left. As they were leaving, I did too. I caught a bus back to Baguio and went home. The PCs were looking for someone that day; and so was the Lord.

Now, back to the house in San Carlos City—about three weeks after moving into the upstairs, the Lord provided another place to live. The owner of the house across the street said that they would rent their house to me. It was a nice house—large, only a few years old, and well maintained. And they left their furniture for me to use. The house had a large kitchen, so I was able to move the refrigerator in.

Ministry in the Public Schools

Our ministry into public schools started soon after I started our school. On Sunday evenings, Pastor Cruz and his family held church services in a small community called Maningding, where he was also building a church. Sometimes, he would invite me to go along.

One town we had to drive through on our trip to Maningding was Santa Barbara. One evening, as we drove back through Santa Barbara on our way home, I saw many small groups of young people standing along the street. I prayed and asked the Lord to please help us reach them.

A few days later, the principal of Santa Barbara High School, who had been one of Pastor Cruz's teachers, saw him somewhere. As they talked, she asked him to come to her school and minister to the students. When he got back to San Carlos, he asked if I would go to the school and do the ministering. About the same time, one of our students was at the school witnessing to some students. While he was there, someone also asked him to come to the school and minister to the students. He then came and asked me if I would come to the school and do the ministering. We now had two invitations to minister to many of the young people I had seen that evening.

In the Philippines, they have a flag-raising ceremony each Monday morning before school begins. Our time to minister would be after that ceremony and before the classes started. The arrangements were for a teen singing group from the church to minister and then for me to speak. I would have ten to fifteen minutes to speak. It wasn't very long, but I was thankful for anything they would give us.

We arrived early, and they took us into the principal's office. She was very polite and courteous, but I could tell she was having second thoughts. She began to talk about getting permission from the provincial administration. The guidance counselor and some others began coming into her office and talking to her about what we were planning to do. Some seemed to be for it, and some against it. As this went on, the flag-raising ceremony was held, and the students went to their first class. It looked like we had lost our opportunity. It was amazing, but I had a great peace. I said little, just sitting there and waiting as the controversy raged on.

Then all of a sudden, the controversy was over, and they said that since we were there, they would let us speak to the students. They dismissed them from their classes and took them to the gymnasium. I asked how much time we had, and they said to take all the time we needed.

We were told there were about 800 students. After we ministered the Gospel, I prayed with them to receive Jesus. I did not see anyone who was not praying.

The Lord knew the flag-raising ceremony was not the proper place to minister, and He had provided what we needed. We would not have had enough time to do what we needed and wanted to do.

Patience Is Rewarded

Often, when things don't seem to be working out or when we are having problems, we get discouraged. However, if we are patient and trust the Lord, these may be times of some of our greatest accomplishments.

One afternoon, I was going to schools distributing tracts. To get to one of the schools outside of San Fernando, I had to leave the highway and walk across a large, dry rice field. After I was finished in the classrooms, I went to the principal's office to thank him and get some things I had left there. He told me that they were building a new building, and he was waiting for some people from city hall to come and talk to him and issue permits.

Then, I walked to the highway and caught a ride into San Fernando to go to some schools there. The first school I went to would not permit me to come in, which was unusual.

Unfortunately, my conversation with the principal ended up being an argument. I do not purposely confront those who try to hinder the ministry of the Gospel, but those children are God's. He created them for Himself, and no one has the right to restrict Him from ministering to them.

> Then they brought young children to Him, that He might touch them; but the disciples rebuked those who brought them.
> But when Jesus saw it, He was greatly displeased and said to them, Let the little children come to Me, and do not forbid them; for of such is the kingdom of God. (Mark 10:13–14)

This occurrence is the only place in the Gospels that says that Jesus was greatly displeased; all other accounts of things He was displeased with only say that He was displeased.

Then I realized I had left my posters at the first school. At that time, when I went to distribute tracts in the schools, I would either leave or put up a Christian poster with scripture in each classroom. These are beautifully colored 24-inch by 36-inch posters, which have scripture printed on them.

I was upset because of my confrontation with the second principal, and then what irritated me even more was that I had forgotten those posters and would have to go all of the way back there to get them.

After I left the second school, I went to another. It was a large elementary, and they let me right in. After I had finished distributing tracts there, I walked to a place where I could catch a ride and then went back to the first school. Of course, I had to walk across that hot, dusty rice field again. I did that often, but that day I was upset, hot, and tired.

An Unexpected Bonus

When I got there, the people from city hall were there. The principal introduced me to them, and we had a short, pleasant visit. There were two men and two women.

In my shoulder bag, with the tracts, were copies of the first lesson of the correspondence course that was given to us by Christ for Greater Manila. I carried them with me and gave them to those the Lord directed me to. Our ministry name and address were stamped on the lesson. When a person finished the first lesson and sent it back, we would correct it and send it back with their certificate of completion and the next lesson. In that package, we also sent them tracts and a few of the first lesson booklets to share with family and friends. We took every opportunity to share the Gospel with everyone we could.

We also often made the correspondence course available to the school children when we went into the classrooms and taught and gave Bibles to them.

I had a leading from the Lord in my heart to give each of the people from city hall a copy of the first lesson, which I did. One woman ended up being one of the most outstanding correspondence school students we ever had, and I know in my spirit that she had a great influence on the others at city hall. What seemed like a bad afternoon ended up being a very good afternoon!

I went back to the second school the next morning and gave tracts to the students and teachers as they arrived. We should never let the devil stop us from finishing the work the Lord has given us to do. Nothing was lost that afternoon, and a lot was gained.

We also had other ministries in Santa Barbara. One day, three men flew over that and another town and dropped our tracts. Also, one of the students in the second class lived in that area, and as he traveled back and forth to school every day, he would distribute tracts to the people who rode in the jeepneys and buses. Another ministry was able to go into that school later and preach the Gospel again.

About six years later, I had to make a transportation transfer very close to Santa Barbara. While I was there, I could sense a wonderful Spirit of peace.

The Fight of Faith

Things progressed smoothly until a woman associated with the ministry noticed an open sore on her ankle. Day by day, it got worse, until she thought she could see the bone.

She became weaker and started to get sick. She was still teaching and ministering, but then bleeding started. It is so hot there, and there are so many infections that we became concerned. If gangrene set in, she could be dead within days.

Weeks turned into a month, and she kept getting worse. We all prayed for her, but she didn't get any better. We were getting worn out. We should have been resting in faith, but we were working hard.

One day, she told the students that if she died, God had not missed it; she had. By this, you can see things were getting serious, because we were losing hope. Another reason for her statement was to counter an attitude that was going through the churches about that time.

A woman who was a teacher at another Bible school had died of some disease. She was known to be a good woman and strong in God's Word. She had served the Lord for many years, but she had died. Because of this, many were questioning whether God would heal everyone. Even our students were fighting doubt and had talked to me about this woman's death.

I had been invited to minister one Sunday at a church in another town. I ministered on the fact that Jesus is always more than enough. I took all the healings and deliverances I could think of in the Gospels and showed that in every case, Jesus was more than enough to overcome the problem.

After that, the woman improved, and within about two weeks, she was perfectly well.

It was then I received a letter from my mother. She wanted to know, "What is wrong over there?" From my mother's letter, I learned that about the time the woman began to recover, the staff of one of the churches in Tulsa gathered for their morning devotions; and as they prayed, the Lord spoke and told them to pray for us.

Even though I did not attend that church, the pastor, missions director, and some of the staff knew me. They prayed for us, and after their devotion time was over, the missions director called my mother and asked what was wrong. He told her what had happened and about the urgency in the Lord's instructions. Of course, she didn't know what was wrong, so she wrote the letter.

Because they knew me, they were able to pray for us, but one of the things that is so wonderful about the Lord is that even though we may not know someone and what their needs are, God has made provision for us to pray for them by the Holy Spirit. After we receive Jesus and the gift of eternal life through Him, God wants to give us the gift of the Holy Spirit.

> And they were all filled with the Holy Spirit and began to speak with other tongues, as the Spirit gave them utterance. (Acts 2:4)

> Then Philip went down to the city of Samaria and preached Christ to them.
> But when they believed Philip as he preached the things concerning the kingdom of God and the Name of Jesus Christ, both men and women were baptized.

Now when the apostles who were at Jerusalem heard that Samaria had received the word of God, they sent Peter and John to them, who, when they had come down, prayed for them that they might receive the Holy Spirit.

For as yet He had fallen upon none of them. They had only been baptized in the name of the Lord Jesus.

Then they laid hands on them, and they received the Holy Spirit.

Now when Simon saw that through the laying on of the apostles' hands the Holy Spirit was given, he offered them money, saying, "Give me this power also, that anyone on whom I lay hands may receive the Holy Spirit."

But Peter said to him, "Your money perish with you, because you thought that the gift of God could be purchased with money!" (Acts 8:5,12,14–20)

And it happened, while Apollos was at Corinth, that Paul, having passed through the upper regions, came to Ephesus. And finding some disciples he said to them, "Did you receive the Holy Spirit when you believed?" And they said to him, "We have not so much as heard whether there is a Holy Spirit."

And he said to them, "Into what then were you baptized?" And they said, "Into John's baptism."

Then Paul said, "John indeed baptized with a baptism of repentance, saying to the people that they should believe on Him who would come after Him, that is, on Christ Jesus."

When they heard this, they were baptized in the name of the Lord Jesus.

> And when Paul had laid hands on them, the Holy Spirit came upon them, and they spoke with tongues and prophesied. (Acts 19:1–6)

Now, with the Holy Spirit's help, we are able to pray for others according to the will of God.

> Likewise, the Spirit also helps in our weaknesses.
> For we do not know what we should pray for as we ought, but the Spirit Himself makes intercession for us with groanings which cannot be uttered.
> Now He who searches the hearts knows what the mind of the Spirit is, because He makes intercession for the saints according to the will of God. (Romans 8:26–27)

> Praying always with all prayer and supplication in the Spirit, being watchful to this end with all perseverance and supplication for all the saints. (Ephesians 6:18)

When you see the term "saint" in the Bible, it is talking about a Christian. One of the names that God calls Christians in the Bible is saints. To God, every Christian is a saint.

I also suggest that you read what Jesus had to say in Mark 16:15–18 and what the Bible says in 1 Corinthians 14.

My Big Mistake

Then, I made a big mistake. Actually, I disobeyed God.

> This is love, that we walk according to His commandments. This is the commandment,

that as you have heard from the beginning, you should walk in it.

For many deceivers have gone out into the world who do not confess Jesus Christ as coming in the flesh. This is a deceiver and an antichrist.

Look to yourselves, that we do not lose those things we worked for, but that we may receive a full reward.

Whoever transgresses and does not abide in the doctrine of Christ does not have God. He who abides in the doctrine of Christ has both the Father and the Son.

If anyone comes to you and does not bring this doctrine, do not receive him into your house nor greet him; for he who greets him shares in his evil deeds. (2 John 6–11)

As the students were going out ministering, a Bible-based cult from the United States began trying to hinder them, arguing with them and trying to put doubt in their minds about the truth of the Gospel. One day during class, two of them came to the door and wanted to talk. I should have taken the advice I had given the students. I told them to never hold discussions with or listen to anyone who tried to talk to them about any other doctrine or religion. I told them to speak only to those who would hear the Gospel of Jesus Christ. I told them that we had been sent to preach the Gospel of Jesus Christ, not to listen to the doctrines or beliefs of anyone else.

This was where I disobeyed God. Since we were in class, and it was just before lunch, I told those people I could not talk to them then; they would have to come back later. They said that they would come back on a certain day at a certain time. On the day these two women came back, a local pastor came to visit, and he talked to them with me. What a bad testimony I gave him.

As we talked to them, they said that they read their bibles twice a year. (As you noticed, I did not capitalize the word "bibles" in the last sentence. The reason is because their founder took a King James

Version Bible and changed it to say what he wanted it to say. It is no longer the true scriptures given to us by God, so it is not worthy of capitalization.) One of the woman told us that she had gone out witnessing a certain number of hours each week for the last twenty years, yet as we tried to show her that we are saved by grace through faith in Ephesians 2:8–9, she could not find the book of Ephesians. In fact, she acted and responded to us as if she had never heard of the book of Ephesians.

As soon as we started to talk to them about the Gospel, the second woman fell asleep and slept until the woman talking to us was ready to leave. The devil had those people in such bondage, and he wanted to enslave us too if we would have given him the opportunity.

Another Mistake

Unfortunately, I did not learn my lesson then. When I left San Carlos City, I moved to Dagupan City and started the next school. On the day one of the classes went downtown to preach in the street, a Caucasian who is not from the United States and his Filipino wife walked by and spoke to us. He said, "Don't forget baptism." I talked to them and gave them my address.

Late one morning, they came to visit, the man, his wife, and their little boy, one of the nicest families I have ever met. It was close to lunch, so I invited them to stay. As we ate and talked, he told me that they didn't believe that Jesus is God.

> Who is a liar but he that denies that Jesus is the Christ? He is antichrist who denies the Father and the Son. (1 John 2:22)

There was the spirit of Antichrist sitting at my table! Good manners or not, I should have immediately asked them to leave my house. It is better to obey God than be concerned about what men think. But I didn't. In fact, I thought I would get them saved.

"For do I now persuade men, or God? Or do I seek to please men? For if I still please men, I would not be a bondservant of Christ" (Galatians 1:10).

We are fellow workers with the Lord, but He does not violate His own Word. If we disobey His Word as we attempt to minister, He will have no part or pleasure in it.

The Damage Was Done

I wanted to talk more to them, but since I was teaching that afternoon, I asked if they would come back on Saturday. They came back and spent most of the day. I did most of the talking; he actually talked very little. He didn't have to talk very much, because the damage had been done. I had received that spirit into the house and greeted him.

> For he who greets him shares in his evil deeds. (2 John 11)

The couple and their little boy left that evening. The next day was Sunday and all was well, but on Monday morning, as I started my devotions, strong thoughts that attempted to block out the truth began coming to me—thoughts that Jesus is not God. I know that if we don't believe that Jesus is God, there is no salvation, so I frantically began to read and confess the scriptures that prove the divinity of Jesus.

After a couple of weeks of this reading and confession of the scriptures, the thoughts left. Then, I prayed and asked the Lord how this had happened. He said sternly to me, "What is the first commandment?" I said, "You shall love the Lord your God with all your heart, with all your soul, with all your mind, and with all your strength [Mark 12:30]."

Then, He said, "What is the second commandment?" I said, "You shall love your neighbor as yourself [Mark 12:31]."

Then, He said, "What is the first commandment?" I answered Him as before.

Then, He said, "What is the second commandment?" Again, I answered Him as before.

He took me over those same scriptures again and again. Then, the Lord said, "I said, 'Don't receive them into your house.'" I said, "But Lord, I was trying to get them saved." He said, "What is the first commandment?" And I repeated the first commandment to Him. Then, He said, "You love Me first. I said, Don't receive them into your house."

The Bible also says in 2 Corinthians 8:5, "And not only as we had hoped, but they first gave themselves to the Lord, and then to us by the will of God."

Dealing with Cults

I understood what He was teaching me. Jesus said, "If you love Me, keep My commandments" [John 14:15]." It is a lack of love, faith, and respect for the Lord to try to do or accomplish anything—even something for Him—in disobedience to Him. Love for Him is demonstrated by our obedience to His Word.

It is foolishness to disobey God, and it can bring great harm. After all, the reason He has called us into the ministry is to do His will.

> As they ministered to the Lord and fasted, the Holy Spirit said, "Now separate to Me Barnabas and Saul for the work to which I have called them."
>
> Then, having fasted and prayed, and laid hands on them, they sent them away.
>
> So, being sent out by the Holy Spirit, they went down to Seleucia, and from there they sailed to Cyprus. (Acts 13:2–4)

In the same chapter, in verse 22, the Lord said, "I have found David the son of Jesse, a man after My own heart, who will do all My will."

Someone will say to me, "I thought we were to go into all the world and preach the Gospel to every creature. How, then, can we preach to the cults?"

The Lord is perfect in all His ways. He has ways to preach the Gospel to those who are oppressed or possessed by the spirit of the Antichrist, and it will not violate His Word. This is the way He chose to minister to that cult in San Carlos City.

A Better Way

One afternoon, two Christian women were sitting and talking in the home of one of them. As they sat there visiting, a lady came in carrying her little daughter. She was taking her home from the hospital after several days of unsuccessful treatment. The doctors had decided there was nothing they could do for her and sent her home. Actually, they were sending her home to die.

The mother looked exhausted, and her eyes were empty with grief and hopelessness. She had nowhere else to go to get help. The hostess explained the situation to her visitor and said, "She is a (and then she named the cult that had come to visit me, the day the pastor was there)."

The visitor told the mother that Jesus would heal her sick child even though she wasn't born-again. But she explained that it would be better if she would receive Jesus and be born again. The woman consented, so she shared the Gospel with her and prayed with her to receive Jesus.

Then, the mother laid the little girl on her lap. The visitor said that when she laid her hand on her forehead, the child was burning up with fever. As she cursed the sickness in the Name of Jesus, the child was instantly healed! She said that as she laid her hand on her feet a split-second later, they were already cool.

Then the visitor, the mother, and the little girl left to go home. The three of them walked down the alley to the street. Across the street from the end of the alley was the headquarters of this cult. On the front porch was a group of women. What a testimony! Most of them probably knew the situation and the hopelessness of it. What a blessing to see God reveal His love and power to them and the truth of the Gospel of Jesus. No one could deny Jesus or what He had done.

> For the Son of Man did not come to destroy men's lives but to save them. (Luke 9:56)

He drew that mother and her daughter there that day to save them. Then, He drew those other women there so they and their families could be saved.

> Then Jesus said to him, "Unless you people see signs and wonders, you will by no means believe." (John 4:48)

No matter who people are or what they've done, God loves them and does what is needed to save them.

The devil will never be able to refute what the Lord has done. That little girl is a living testimony. My disobeying His Word and bringing the cult members into the house did not accomplish anything for the Lord or them. God has His plans and ways of accomplishing them. And they never include our disobedience to His Word.

This is a way the Lord did basically the same thing through me.

When the faucet in my kitchen was not operating properly, I told the land lady, and she contacted a plumber. On the morning the plumber came, I was at home and walked out into the kitchen to see how he was doing.

I noticed he had placed a magazine that was printed by another Bible-based cult on the drain board. This cult also teaches that Jesus is not God. Printed on the cover of the magazine was this question:

Is Jesus God? I asked him if he wanted to know the answer to that question, and he said yes.

I got my Bible and showed him from the scriptures that Jesus is God. Then, I ministered the Gospel to him and asked if he would like to pray and receive Jesus. He said that he would. When he left there, he was saved.

What happened that morning worked, because I did not violate God's Word. I did not invite him into the house, and I did not know what he believed, but when the opportunity to minister the Gospel presented itself, I used it.

The High Price of Disobedience

There was another way I was disobedient to the Lord's instructions. As I mentioned earlier, when I went to the missions director of my Bible school, he told me that he believed the Lord was saying for me not to become associated with any other ministry. And I said that I had a strong witness that what he said was from God. We are never moved by what any man says apart from God.

When I went back to San Carlos City to spend the weekend with Pastor Cruz and his family, I told him I could in no way be associated with his church.

Even though I had moved out of their house soon after school started, I still continued to be involved with his ministry, ministering in the church from time to time, teaching Church sponsored Bible studies, and teaching the Youth Group. At the time, everything seemed fine, but I saw the results of my disobedience a couple of days before I left San Carlos City.

On that day, I went to a restaurant and had lunch. It was a nice restaurant with a large meeting room in the back. As I was eating lunch, the wife of the owner came by to introduce herself and visit. She said that she was a member of the large cathedral in town. I could tell she was a wealthy and influential woman.

She said that their group had wanted to invite me to speak, but since I was part of Pastor Cruz's church, which they considered to be

Protestant, they had not invited me. I assured her I was not part of that church, nor was I a Protestant; I was a Christian. "Protestant" is man's term. But it was too late. I had lost my opportunity to minister to the religious leaders in that town.

Next, I went to an appliance store to buy a refrigerator for the duplex I was moving to in Dagupan. The store owner told me that he was the past president of the local Rotary Club, and they had wanted to invite me to come and speak to their luncheon, but since I was part of Pastor Cruz's church and since the club members were not, they could not invite me. There had been my opportunity to minister to the business leaders of that town. It also had been lost! I learned it is essential to be absolutely submitted to the Lord and obedient to His Word if we want to fulfill His plans for our lives and ministries.

I always told the students when they go out to minister to never mention the name of their church. No matter how hard anyone tried to find out what church they attended, they were not to tell them. I told them when someone asked what church they attended to tell them they were not there to talk to them about their church, but to tell them about Jesus Christ. Often when people find out that your church is not their church, they will not listen to you.

The students ministered in San Carlos for about five months. They witnessed everywhere—door to door, in the park, and in the hospitals. They preached in the street, the jail, the schools, and so forth. There was a pastor there, a sincere man of God, who told me the students should direct people to his church. I told him no; the Holy Spirit would direct them to where He wanted them to go.

When I arrived in San Carlos City, there were fewer than 200 people attending that pastor's church on Sunday mornings. A couple of weeks after I left San Carlos City, I saw one of the elders of his church in Dagupan. When I asked how the church was doing, he said, "If we could seat them, there would be 2,000 people in the church." God will direct His people where He wants them to go.

The Lord provides!

One day during class, we had a wonderful example of the Lord's provision for the students. I was continually seeing the Lord's desire and ability to provide for our needs. However, the students never really saw the Lord's involvement in this area. Everything that was being provided to the school and students came through me, so they never saw Him in what they received. When they came to school, everything was free, including their room and board.

The Lord spoke to me after the first class had graduated and admonished me in two areas. The first thing He said was, "Some of these students came here believing Me and now they believe you." He wanted me to help them, but as He led and directed.

The object was to train them for ministry, and part of ministry is exercising faith in God for provision. Ministering Christianity is teaching people to believe and trust the living God who delights to perform His Word and bless people.

The second thing He said was that He did not send me there to live like that. While in the United States, I had been convinced by others that I should live and eat like the poor. During the first class, I lost a lot of weight, got sick easily, had little energy, and was not as mentally sharp as I should have been. It is important to understand that our metabolisms are different from people who live in other countries, and theirs is different than ours.

This is what the Lord did to reveal His provision to the students. In the Philippines, most people eat rice each meal. It is a staple of their diet. One morning, there was only enough rice for the students' breakfast. This meant there would be none left for lunch. We had other food and everything else we needed, but we were out of cash. It was the end of the month, and I had already given my check to the man to cash and was waiting for him to bring the money.

About ten thirty that morning, as we were in class, the mailman came and stuck a letter through the window. When I opened it, there was a ten-dollar bill inside. If it had been a check, I could not have cashed it. At noon, I sent one of the students to the market to buy rice.

Being out of rice to those students was like being out of food. By 10:30 a.m., they were probably beginning to think about lunch and what they would eat. That was the only time I came close to running out of food for the students.

That letter had been mailed from the United States about two weeks earlier. It had gone through a number of different mail systems and was handled by many people. The money was sent by someone who didn't know we needed it, but it had been mailed on time.

> Now in the fifth year of Cyrus King of Persia, that the word of the Lord by the mouth of Jeremiah might be fulfilled, the Lord stirred up the spirit of Cyrus king of Persia, so that he made a proclamation throughout all his kingdom, and also put it in writing. (Ezra 1:1)

Praise the Lord! He can stir the spirit of anyone He chooses to do whatever He wants done, so His Word might be fulfilled to us. He is awesome! This time when He met our need, He did it in a unique way—in such a way that He could reveal His love and care to the students. It ministered to them because that rice was really for them, and they knew it.

Touching Others

As we rest in the Lord and don't get anxious or uptight, even when things seem to be going wrong, we give Him wonderful opportunities to reach out through us and touch others.

One day, I had to go to the Manila airport to pick up a package that had been sent from the states. The taxi driver whom I hailed asked if I wanted him to start the meter or let him charge me a flat fee. (He was trying to get to me.) I asked him how much the flat fee would be and he told me 150 pesos. I knew from experience that the cost to the airport should be about 75 pesos, so I told him we would use the meter.

As we were driving along talking, he mentioned that he had a cold and felt terrible. I told him if he would let me, I would lay hands on him and pray, and Jesus would heal him. He agreed, so when we stopped at a red light, I reached forward, put my hand on his shoulder, and prayed. I don't know what happened; but when I prayed, something inside of him seemed like it broke loose; and he broke down, crying profusely.

Since things were going so well, I kept him for the trip back. As we rode, we talked about the Lord and had a good time. Since it was near noon on our trip back, I offered to buy his lunch, and he accepted. Since it was important to him to keep his meter running, the Lord directed me to take him through a McDonald's drive-through window.

We sat in the parking lot and ate and then went to a Christian bookstore, where I bought him a Bible.

There was a Bible study that afternoon at one of the homes in the compound where I was staying. When he dropped me off, I forgot some papers in his taxi, and later, when he found them, he brought them back. He asked some of the women who were entering for the Bible study to give the papers to me. I don't know what he told them, but they were impressed with the way he talked and acted and could tell that the Lord had done something for him. They were very impressed that he brought the papers to me and did not keep them. Everyone was blessed to see what the Lord had done for and in him. No one knew him, but you could see that the Lord's hand was on him.

Sometime later, I was in a taxi picking up Bibles at Christ for Greater Manila (CGM). It was just before noon, so on the way back to where I was staying, I thought I would try to bless this taxi driver the way I had the other one. Since we were near the same McDonald's restaurant, I asked the driver if I could buy his lunch.

In a very unfriendly way, he said no. It was very hot, and we were perspiring, so I thought that I could at least buy him a coke. Since I was paying the meter and it would not cost him anything, I told him to go through McDonald's drive-through window. When he did, I bought both of us a large coke even though I didn't want one myself. After he had finished his coke, he turned to me and said, "Now I have ruined my lunch." One of the things I learned from that is not to try to control people and situations but to be led by the Lord and allow Him to minister. When He does, it works.

Overcoming Feelings

Once, we were invited to a church in another area to hold meetings over the weekend. We were to leave on Friday right after lunch. The enemy was strongly resisting us, and I was not feeling very excited about this weekend of ministry and for no good reason. The reason the devil resists us from ministering is that he knows that when Jesus gets there, he has big problems.

After lunch, as I was getting ready, the students came to the door and said that they had decided not to go. You cannot imagine what an unusual thing that was. Not only did they always obey and show me respect, but also they were quickly becoming trained and disciplined ministers of the Gospel who understand there is never a reason not to go to minister.

In Matthew 8:21–26, we see a similar situation:

> Then another of His disciples said to Him, "Lord, let me first go and bury my father."
>
> But Jesus said to him, "Follow Me, and let the dead bury their own dead."
>
> Now when He got into a boat, His disciples followed Him.
>
> And suddenly a great tempest arose on the sea, so that the boat was covered with the waves. But He was asleep.
>
> Then His disciples came to Him and awoke Him, saying, "Lord, save us! We are perishing!"
>
> But He said to them, "Why are you fearful, you of little faith?" Then He arose and rebuked the winds and the sea, and there was a great calm.

As you continue to read this account, you will find that as they continued on with the Lord, He delivered two demon-possessed men. They were on their way to set the captives free and the devil knew that when he came in contact with Jesus, "the Rock," something would have to move, and it wouldn't be Jesus!

The morning I arrived in Manila, Mrs. Cruz's son, Steve, picked me up at the airport. When we arrived at their home, they took me to my room to rest. As I stepped into the room and closed the door, a tremendous weight came upon me. It felt as if I was being pushed through the floor. All I could do was lean against the wall and say, "Lord Jesus, help." When I said that, whatever it was instantly left.

> I will call upon the Lord, who is worthy to be praised; so shall I be saved from my enemies. (Psalm 18:3)

> He delivered me from my strong enemy, from those who hated me, for they were too strong for me. They confronted me in the day of my calamity, but the Lord was my support. (Psalm 18:17–18)

> You called in trouble, and I delivered you; I answered you in the secret place of thunder. (Psalm 81:7a, b)

The week before I moved to San Carlos City, I became very sick. On the day I left, I woke up feeling terrible; but as the day progressed, I felt better and better; and by the time I arrived, I felt fine. We can never give in to feelings. We must trust the Lord and move ahead, and He will change the feelings.

After I moved to Dagupan, it seemed that every Sunday I was invited to minister, I would wake up with diarrhea. And over there, there are not many public places to use the bathroom after you leave home. And many other times, I would be scheduled to minister and sickness would come against me. But thank God, as I stepped out in faith and moved ahead, He always healed me. I only remember missing one ministry opportunity.

I told the students that we were going and sent them to get ready.

When we arrived, the pastor of the church was sick. The devil was resisting us, but as we continued to look to the Lord, He strengthened us so we could move ahead. They took us to an elementary school near the church and brought all the children outside. We preached the Gospel to them and prayed with them to receive Jesus. Afterward, we went to the homes where we were to stay so we could rest and prepare for the evening service.

At 6:30 p.m., the lights went off. I found out later that sometimes, the electricity in that area goes off for as long as three days. We walked to the church, and when we got there, I prayed and commanded the electricity to come back at seven o'clock. As we stood there in the dark, the pastor began to ring the bell. The bell was a piece of big pipe, which was hung by a rope by the door and was hit with a smaller pipe. As we stood there, the pastor gave me the next bit of news. He told me he didn't think anyone would come. He said that they never had Friday evening services.

All of a sudden, the lights came on. I looked at my watch, and it was exactly seven o'clock. In Jesus's Name, God will do what we command.

> Thus says the Lord, the Holy One of Israel, and his Maker: "Ask Me of things to come concerning My sons; and concerning the work of My hands, you command Me." (Isaiah 45:11)

Acting in Obedience

As we filed into the church to start the service, an elderly man walked in and sat at the back. When we started the service, the congregation consisted of that man and us, the ministers. But then people started coming in. In a little while, the church was full, and they had to open an overflow room. This church was beside MacArthur Highway with houses and small communities scattered along the highway near it. The pastor said that he saw people in church that night he had never seen before.

Some girls said that they didn't know why they had come to the church; they were walking down the highway, and suddenly, they turned into the church. One man said that he had a vision of hell. We had a wonderful time that night. I don't know how many people were saved.

The next morning was to be street preaching by the students. We would walk along the highway until we found a group of people, and one of the students would preach to them.

At one store we went to, fourteen men were sitting outside. They were a bunch of mean-looking guys! Someone later told us that they were ex-convicts, and I believe it. I believe I could see murder in the eyes of some of them.

It was one of the women student's turn to preach, but these men just glared at her, and she gave up. After she stopped, another woman started to preach, and after a while, the men began to break. Tears started to form in some of their eyes.

When she had finished preaching to them and praying with them to receive Jesus, they were totally transformed. I remember some of the men coming to me afterward, holding my hand, and saying, "Thank you, sir. Thank you, sir."

All of those men were born again that morning. Only the saving power of the Gospel can change men's lives!

That was the afternoon our students took the students from the other Bible school out witnessing door to door. More than 130 were saved during that door-to-door ministry. I have no idea how many were saved during the entire weekend. When the Lord called me into the ministry, He said, "Go into all the world and preach the Gospel to every creature." And He is fulfilling that call.

CHAPTER 5

The Power of Tracts

In early 1982, a ministry from the United States came to the Philippines to hold meetings in Manila. Part of their street ministry was distributing tracts. Afterward, they gave me a box of tracts they had left over, so I took them back to San Carlos City and gave them to others to distribute. Some of these people came back and asked for more. So, on my next two trips to Manila, I stopped at this ministry to pick up some more tracts. On both times, they were out of stock, but told me they planned to have more printed.

One morning shortly after this, as I was waking up, the Lord spoke to me and said, "I want you to write a tract." I thought, *I don't know what to write*. He said, "I'll tell you what to write." I have inserted a copy of the tract at the end of this book.

When He said, "Go into all the world and preach the Gospel to every creature," I didn't pick up on the "into all the world" part, only the "go and preach the Gospel" part. Since 1982, the tract has been translated into all of the major Philippine dialects. And some individuals have taken it or sent it to other parts of the world. Thus, I am seeing a partial fulfillment of His charge to me.

Success with Moslems

Once, a ministry in western Mindanao, which was predominantly Moslem, asked for tracts. The pastor and his wife had been

Moslems but were now born-again. When we received their request, we sent them 30,000. The report we got back later, after they had given out the tracts, was that they were having a problem finding places big enough, and fast enough, to accommodate all the people who were being born again!

This tract is God's Word; therefore, He can work with it. I have been told that it is quite effective with cults and false religions. Someone else told me that one of the reasons it is so effective is because it is written like a story, and everyone likes a story. The Bible is a book of stories. But the major reason it is effective is because it is God's Word, anointed by the Holy Spirit.

We received a funny side note from the people who were distributing the tracts in that Moslem area.

In the Philippines, many of the political candidates print and hand out handbills during campaign time before elections. Many Moslems in that area have taken Middle Eastern names so when they gave a tract to one man he thought it was a political handbill. When he looked at it and saw the name, he said, "Jesus Sa-eed, who is Jesus Sa-eed?"

Another ministry opportunity happened as I was walking past what appeared to be a pile of tree limbs about four feet high and eight feet across with some strips of plastic stretched over the top.

As I looked closer at it, I could see there was an opening in the side facing the street. As I continued to look, I realized that people were living in there. I found out that a whole family lived there. It was terrible!

It was the rainy season, and that little bit of plastic did not cover the whole top and could not keep all the rain out. It looked like the floor had a few boards laid on it, but the rest was mud. The family who lived there consisted of a father, a mother, and two small boys. The father was crippled and could not work, so the mother sold little bags of peanuts in the market to make a living. She barely made enough for them to live.

When I saw their situation, I decided to build a house for them. It would be a one-room nipa house, which is common there and would be adequate for their needs. I found out what the cost would be and appointed Daniel, one of the students, to be the construction manager. I gave him the money, and he made the arrangements for the materials and the workmen and oversaw the job. What a blessing to walk by later and see their new home. It cost about one hundred dollars, and that was one hundred dollars well spent. Thank God for all of the opportunities that He gives us, the Body of Christ, to help and minister to others.

After the house was completed and they had moved in, Daniel went by and shared the Gospel with them, and the parents were born-again. Sometime later, the woman came by and gave us three or four small bananas. To them, this gift was probably a great sacrifice, and to me, it was priceless!

We built that house in 1982. In 2000, I was visiting Daniel at his church. And as we sat in his office talking, all of a sudden, he snapped his fingers and said, "Do you remember that family that we built the house for?" When I said, "Yes." He said, "They began to go to church, the father got healed, the children got born-again, and today, the whole family is on fire and serving the Lord."

There was also an older woman in her seventies whose house was starting to fall over. She was taking care of two of her small grandchildren and didn't have enough money for repairs.

In the Philippines, because of the flooding monsoon rains, many of the nipa houses are built on heavy bamboo posts. The posts supporting this woman's house were rotting away, and the house was in danger of falling over. The women's group in Pastor Cruz's church was afraid to meet there anymore. They said that when they went into the house, it would move and sway. Praise the Lord! I was able to buy new posts for her. They cost between $20 and $30. Sometimes, it doesn't cost much to be a great help to someone!

There are so many wonderful things you see and hear in the Body of Christ. Here's some that blessed me.

A woman who was born again in Pastor Cruz's church wanted to be baptized. Since most baptisms in that area are held at the beach, she would have to go there. On the day they were having a baptismal service, she and a number of her children began walking to the beach, which is about twenty or twenty-five miles from where they lived.

It would have cost less than five dollars for all of them to go round trip by public transportation, but they didn't have the money. After they had walked a long way, someone from the church picked them up and gave them a ride.

What blessed me was that this woman loved the Lord so much that she was willing to do whatever was necessary to fulfill His Word, and she was teaching her children to do the same.

The next one was the time I was invited to Bacolod City to do some ministry. While there some of the brethren told me this story. They said that they live on an island without electricity. Since they wanted to be able to have music in their church services, they decided to pray and ask the Lord for a Boombox. One day a large ocean-going ship, from Korea, stopped beside their island. As they watched, the ship let down a small boat and some men from the ship, boarded the small boat and headed for their island. They said that they ran down to the beach to see what they were doing. When the men arrived, they asked where the Christians on that island were. When they told them that they were the Christians on that island, they gave them a Boombox, went back to the ship, and the ship left. There is nothing too difficult for our God!

Something else the Lord did to provide for a woman and her seven children happened in 1986. Remy was a student in our second class, who came to school the days that she could. She had a job, selling household cleaning supplies door to door. She was a registered nurse and could have gone abroad to work and make more money, but she thought she should stay in the Philippines and "bring up her children in the training and admonition of the Lord," according to God's word.

When I returned to the Philippines, I was asked to teach a prayer seminar in Dagupan City. Remy came to the seminar, and after it was over, we had an opportunity to visit. As we talked, the Lord said to me, "Take Remy and her children to Baguio." When I asked her if she wanted to move there, she enthusiastically said, "Yes."

She and the children were able to move into a nice apartment, and they began their lives there. A short time after that, I was given a gift of over 20,000 pesos. After I took out the tithe and some money that I needed, I had about 15,000 pesos left. I prayed and asked the Lord what He wanted me to do with the remaining money, and He said, "Buy clothes for Remy and her children." The clothes they had brought from Dagupan were warm weather clothes, and Baguio, being in the mountains, was quite cold.

One day, a few months later, I asked Remy how she was doing, and she said that she was having a difficult time financially. The company she worked for increased the percentage rate of commission they paid as a person reached certain plateaus of gross sales. Because she had left Dagupan and had started selling in Baguio, they had set her commission back to the lowest rate, and she would have to build it back up again.

A few days later, I was back in Manila. While there, some friends invited me to go with them to see a famous American basketball player, who was speaking at a local church.

After he finished speaking, the pastor of the church came to the pulpit and asked if anyone wanted prayer for anything to come forward. Tony, one of my friends, said to me, "I'm going forward for prayer." Even though I didn't need prayer for anything, I decided to go too.

The pastor had some of the church members form a line at the front, and we were to go to one of them for prayer. When the man I went to asked what I wanted him to pray for, I told him I was a missionary and asked him to pray for my ministry. After he prayed, we began to talk. As we talked, I asked him what his profession was, and he told me he was the president of a certain company. When he said the name of the company, I knew it was the company that Remy worked for. Isn't our God Great! When he said that, I told him about Remy and her children and how she was struggling financially because of the low commission rate that she had been moved back to. He gave me his card and asked me if I could bring her to see him the next Wednesday afternoon at three o'clock. After arriving at his office, they talked for a while, and as we were finishing our visit, he told her that he would raise her commission back to where it had been before. "Then Job answered the Lord and said, 'I know that You can do everything, and that no purpose of Yours can be withheld from You'" (Job 42:1–2).

I am writing this in July of 2012 as I prepare this book for the second printing. Today, Remy's office is on Session Road in Baguio City. Session Road is the most expensive real estate in Baguio. All of her seven children have attended college. Six of them have finished and earned their degrees. Remy's oldest daughter, Bing, is married to a pediatrician; and the second daughter, Mars, is married to a Philippine Army doctor. She was close to thirty years of age when she married. She was waiting for God's best, and He didn't disappoint her. It takes a long time for a young man to be educated and go through all of the training necessary to become a doctor. "Imitate those who through faith and patience inherit the promises" (Hebrews 6:12b).

I was invited to Mars's wedding, and the whole wedding and reception afterward were totally centered on the Lord. And her husband is a fine, godly young man.

I asked Remy if I could include what the Lord has done for her and her children in this book. She responded, "You can tell the whole world what the Lord has done for my children and me." So we will.

There is one more thing the Lord used Remy's situation to do. Remy would sell cleaning supplies during the week, and then on the weekends, she would have to travel to Manila to pick up the supplies to fill her orders. A round trip would take at least twelve hours. Since I was going to Manila each weekend for ministry, I told her I would take her order to the company, get it filled, and bring it back to her.

When I arrived for the first time, I found that the place I was to get the orders filled was a large warehouse. When I arrived, there were many other salespeople there picking up orders also. As I talked to the employees, I found out that many of the people who worked and sold for that company were Christians. Looking for every opportunity to share the Gospel, I asked if they would like to have free tracts. They enthusiastically said, "Yes." Every week for a period of time, I brought tracts to them. Many of the salespeople distributed the tracts while they went about selling their products.

There is something else I believe you will enjoy reading about Remy's children. When Remy was attending school, she and her children would come to dinner from time to time. (Alpha, the youngest, was five, and Jose, the oldest, was sixteen.) After dinner, we would usually have a Bible study. Remy told me that when the younger ones found out that you could ask God for things and He would give them to you, the younger ones got together and decided they would pray and ask for bicycles. A little while later, the Lord spoke to Larry, my next-door neighbor, and told him to buy bicycles for them. And, of course, he did. You will read more about Larry later on in this book.

Then, there was the time I was invited to be the afternoon speaker at an area rally for a certain denomination. I didn't know it, but that group believed and taught that we should thank God for everything. There were a number of pastors and church officials there as well as members of their congregations. That afternoon, I spoke on 1 Thessalonians 5:18:

> In everything give thanks; for this is the will of God in Christ Jesus for you.

I taught them that the scripture does not mean that everything that comes into our lives is from God. I taught them James 1:16–17, "Do not be deceived, my beloved brethren. Every good gift and every perfect gift is from above, and comes down from the Father of lights, with whom there is no variation or shadow of turning." If what comes into our lives or the lives of others is not good and perfect, it is not from God.

> The thief does not come except to steal, and to kill, and to destroy. I have come that they may have life, and that they may have it more abundantly. (John 10:10)

If what comes to us steals, kills, or destroys, it is not from God. Because some people have seen God working in terrible circumstances and bringing some good out of them, they think He brought or caused the problem. In Romans 8:28, the Bible clearly says, "And we know that all things work together for good to those who love God, to those who are the called according to His purpose."

In the first two chapters of the book of Exodus, we see the Egyptians throwing Hebrew babies into the river to kill them. This was not God's work in the lives of His people, but the devil's through the pharaoh. However, God turned it around and blessed Moses's parents through it.

This account shows how great our God is. Moses was God's deliverer, and the pharaoh was Satan's destroyer. God placed His deliverer in the house of Satan's destroyer, and Satan's destroyer raised God's deliverer for Him, and Satan couldn't do anything about it! If God is for us, who can be against us?

1 Thessalonians 5:18 does not teach us to thank God for everything; it teaches us to thank God in everything. Thank Him for what? Thank Him, as it says in Romans 8:28, that He is working in that circumstance for our good. That is why we so often see some good come out of bad things.

I know the Bible says in Ephesians 5:20, "Giving thanks always for all things to God the Father in the name of our Lord Jesus Christ." If your Aunt Susan gives you a birthday present, you don't thank your Uncle Frank for it. If we have that much understanding about natural things, we should apply that same understanding to spiritual things. We are to give thanks to God for all the things that come from Him.

We are not to thank God for what does not come from Him. I taught this group that we should judge all things by God's Word and deal with them appropriately. If they are from God, they will bless and prosper us, and we should appreciate them and thank God for them. But if they come from the thief, they are not His will for our lives, and we should resist them through His Word and faith. If we allow sin or anything else that comes from Satan to come into or remain in our lives, it will ultimately steal, kill, and destroy. That is why many Christian's lives and homes are destroyed; they are thanking God for what the devil is bringing to them and not getting rid of it!

I was very glad I had that opportunity to share the truth with them; however, they did not invite me back.

I had scheduled the first class term to last for six months. A few days before the term was to end, I went to Manila to renew my visa. As I was riding along in the bus, the Lord said, "I want you to extend school six more weeks. The students have been working hard, and I want them to bring in the harvest."

I said, "Lord, You will have to talk to the students and the owners of the house." The students were getting ready to leave school, and I knew they had made plans. I also knew the owners of the house wanted it back. A few days earlier, the owner had talked to me and asked when I would be ready to vacate the house. I told her when I would be leaving, and she seemed satisfied. She told me that since the rainy season was starting, they wanted their house back. It would be a better place to go through the monsoons than where they were staying.

Bringing in the Harvest

The next morning as we began class, I told the students what the Lord had said. When I asked how many wanted to extend school for six more weeks to bring in the harvest, they all enthusiastically raised their hands.

At noon, I talked to the landlord and asked if I could keep the house for six more weeks. She happily said yes. She talked and acted as if she had not talked to me at all about wanting their house back. So we were excitedly set to bring in the harvest.

The classroom schedule was from 8 a.m. until about noon. After lunch, at about 1:30, the students went out to minister, returning between 4:30 p.m. and 5:00 p.m. As I mentioned earlier, when I took the students out to preach in the street on the first day of their outside ministry, I had received so much attention that I realized I could no longer go out with them.

Since I could not go along and help them, I talked with them before they left about where they would go and what they would do.

When they got back, I talked to them again about what they had done and the results.

One thing I told them to do was count how many people were born again. This way, I could judge how effectively they were ministering. If they had a prolonged slump, I would know we had a problem. The three teams of two each were averaging about fifty people being born again each day.

As a side note that I think is relative to what the Lord was doing at that time, I would like to tell you about a problem I had with one of the women students. She was a good student, and people told me she was a good teacher after she left school.

As we were nearing the end of school, she became more and more rebellious and would not submit to my authority, particularly when it came to outside ministry.

Because she was much older than the other students, I was concerned about her negative attitude on them. I was also concerned they would follow her lead and not my instructions when they went out each afternoon to minister. I realized that unless this situation was corrected, the students' education would be harmed, and our outside ministry would be hindered, if not destroyed. So I talked to her repeatedly, but she refused to submit. Finally, she quit school. It was a sad situation because she was within two weeks of graduation. (This was before the Lord told us to extend school.)

The Lord Provides

When she left, Digna had no one to go out with her, but we found that the provision for our need had already been provided. When Juhn moved to the school, he brought Ligaya, his wife. Ligaya's father was the pastor of a church on another island, so she had been raised in the church and was a good minister herself. She had attended the classes that she could, so she was familiar with what I was teaching and doing. (By the way, Juhn and Ligaya had their first baby while they were at school and named him Habakkuk.)

Ligaya consented to go out with Digna, so our three teams of two each were preserved.

Earlier in the year, one of the male students had quit, but shortly afterward, a young man named Manolo came and asked if he could attend classes. Even though he could not graduate because he did not attend the full class term, he was well enough prepared to participate in the outside ministry. Both students we lost were replaced by qualified people, and we were able to finish what the Lord had called us to do.

We extended the school and started to bring in the harvest. At the end of three weeks, I was disappointed. About a hundred people a day were being born again. That was twice as many as had been born again before, but it was not what I called a harvest. On the third weekend, I prayed, "Lord, I don't understand. I thought You said we were going to bring in the harvest." He responded, "How many do you want Me to save?"

I have never heard that question before, so on Monday morning, I said to the students, "The Lord asked me how many do we want Him to save?" There was silence. No one said anything.

I reminded them that approximately 1,500 had been saved the first three weeks and asked, "How many of you think the Lord can save twice that many [3,000] in the next three weeks?" All the students put up their hands. We were agreed, so we prayed and asked the Lord to save 3,000 people in the next three weeks. That day, Daniel made a sign that read, "3,000 saved in three weeks" and hung it on the wall for us to look at to reinforce our vision.

A Great Harvest

By Tuesday evening of the following week, the Lord had saved the 3,000! That was more than 400 per day. It was great, and we had only used seven of our fifteen ministry days. We still had eight ministry days left. So the next morning, I asked the class, "Since we have our 3,000 saved already, how many do you think the Lord can save in the next eight days?" Again there was silence.

After a while I asked, "How many of you think the Lord can save another 2,000 in the next eight days?" Everyone raised their hands. This meant that the Lord would save 5,000 in the last three weeks. Again, we were agreed, so we prayed and asked the Lord to save 2,000 more in the next eight days. Our sign now read, "5000 saved in three weeks." The students were now going into the schools, and by week's end. we had our additional 2,000 salvations. That was almost 700 a day!

On Monday morning, I said, "Well, we have our 5,000 saved, and one more week to go. How many do you think the Lord can save this week?" Again there was silence, so I said, "How many believe the Lord can save 5,000 this week?" Everybody's hands went up. We prayed for the last time and asked the Lord to save 5,000 that week. On Friday evening, when the students had all reported in, we had our 5,000 saved! That week, they had seven opportunities to go out and minister, because I let them go out on Thursday and Friday mornings also.

When we put together our final total for that six-week period, we actually had more than 12,000 persons born again.

A New Call

Right after I had arrived in San Carlos City, I was talking to someone about Baguio City, a beautiful resort city in the mountains. It was a city full of occult practices. People went there from all over the world for occult healings.

I told this person someone should go there and start a teaching and healing center. Then, when people went there for occult healings, they would have an opportunity for the Lord to minister true healing to them. As I talked to her, it seemed to me that the Lord was saying I would be the one to do it.

Because school was coming to an end, I thought that might be the time I would be going to Baguio. But on a Sunday afternoon trip to Dagupan City, I had a witness from the Lord that I was to move there next.

The famous Blue Beach at Dagupan is where General MacArthur made his Luzon landing.

Once I knew I was going to Dagupan, I started to think about my next home. I decided this time I would write a description of the house and what I wanted it to include and then pray for it.

First, I did not want it to be on a busy street again. In San Carlos City, I lived on a main street that led out of town, and it was loud from early in the morning until late in the evening. The list included three bedrooms, two bathrooms, and screens on the windows. I prayed for my new home and read the list to the Lord.

The last week of school, I went to Dagupan to look for my house. On the first day as I was riding to Dagupan, I saw a three-story building along the highway. It was obviously a commercial building, so I didn't pay much attention to it. However, I did notice a sign that said, "Rental Information call 4285."

Through that building and rental sign, the Lord would lead me to my new residence. Often, this is the way the Lord leads us to His provision. He will lead us to one thing, and through that, He will lead us to what or where He wants us to go. Sometimes, there's one step, and sometimes, there are a number. But at the end of His leading, there's always a wonderful blessing.

I went downtown and walked around, but could find nothing. I asked a pastor if he knew of anything, but he said that he didn't. However, he said that he would look around and also ask the members of his congregation if they knew of anything.

Later, I met a man who owned three clothing stores in Dagupan. He also owned rental property and was a realtor, but he didn't know of anything. He called the other realtors in town, and they didn't know about anything either.

The pastor walked around town with me and then drove me around in his car to try and help me find something. Then the man with the clothing stores also drove me around in his jeep, but we could not find anything either.

On the first day, I walked past that three-story building and looked at it. The first floor was a wholesale television dealership. The second floor was empty, and the third floor was a radio station. I didn't see it before, but as I write this, I realize the vacant second floor was not right for me. It was on the highway and that was something that was on the list that I didn't want.

Day after day, I went to Dagupan and looked for a place, but I could not find anything that suited me. On Friday afternoon, I heard about a place out by the beach, but it was too late to go that day. That night was our graduation dinner, and the next afternoon was the graduation service. After the graduation service, the students would go home. (Just a side note to show how great those students were—there was a school south of San Carlos City they had wanted to evangelize, but they ran out of time and were not able to get it done before school ended. The following week, when school was finished and they had graduated, when they didn't have to, some of the male students came back to San Carlos City to finish the work on their own initiative and at their own expense. Whatever it cost to go and train this caliber of people was well worth it.)

I made arrangements to move back across the street on Saturday before graduation, so the people who owned the house could move back in. Everything was set, except I didn't have my house.

I decided to go back to Dagupan early Saturday morning and look at the house and be home in time for graduation. If the house was not what I needed, I would go to Baguio City on Monday. Maybe I had missed the Lord.

A New House

On Saturday morning, I was in Dagupan early. I went out by the beach and walked around looking for that house, but could not find it. As I walked, I asked the people I saw if they knew about a house for rent, but no one did. I decided that was it; I would go to Baguio City on Monday morning.

As I walked toward downtown Dagupan to get my jeepney back to San Carlos City, I walked past the provincial hospital. As I got in front of the hospital, I could hear that telephone number being repeated over and over inside of me: 4285, 4285, 4285. I went into the pharmacy next door to the hospital and used the public telephone. When I reached into my pocket for the change, I was surprised to find I had the four ten centavo coins that I needed to make the call. Normally, I would not be carrying that many of those small coins. You rarely use them for anything.

When a woman answered that number, I asked her if the second floor of the building on the highway was commercial or residential. She said commercial, so I thanked her and began to hang up. As I was hanging up, I could hear her say something, so I put the receiver back to my ear and asked her what she had said. She asked if I was looking for a place to live, and I told her yes. She asked me if a duplex would be all right, and I said that it would. She said her father-in-law had a duplex for rent and gave me the directions.

When I arrived, the man was waiting and showed me a beautiful, brand-new American duplex no one had lived in before.

> For the eyes of the Lord run to and fro throughout the whole earth, to show Himself strong on behalf of those whose heart is loyal to Him. (2 Chronicles 16:9)

The man who owned the duplex had brought the plans for it from the United States. It was exactly what I had prayed for—and more.

> Now to Him who is able to do exceedingly abundantly above all that we ask or think, according to the power that works in us, to Him be glory in the church by Christ Jesus to all generations, forever and ever. Amen. (Ephesians 3:20–21)

It was off the highway and quiet. It had three upstairs bedrooms. This was important, because the soft, cool evening breezes that made sleeping comfortable were able to flow unrestricted by ground-level vegetation; it also had two bathrooms.

As the owner was showing me the house, I stood there amazed. It had beautiful white terrazzo and marble floors downstairs and hardwood floors upstairs. It was attractively and brightly painted. The owner had built a water tower and dug a well, so the house had water all the time. Because the mayor lived in that area, there were only a few electrical outages in the nearly two years I lived there.

I had forgotten all about screens for the windows. But as we stood in one of the bedrooms talking, the owner said, "Oh, by the way, if you rent this place, I will put screens on all the windows, free." All things are possible with God.

God's Bonuses

Besides all that, there was another bonus. Two American medical students lived next door. One returned to the United States, but my dear Christian Brother Larry stayed and lived next door the whole time I was there. We ate our evening meals together, so we had rich times of fellowship.

As I said before, there was nothing suitable for me to rent in Dagupan and a lot of people were helping me look. Rental property was very hard to find. The driveway back into my new home began at a highway. When I got there to look at the duplex, there was a big sign on the gate post beside the highway advertising the duplex for rent. That sign was right out in the open, you couldn't miss it, and yet they had been trying to rent that place unsuccessfully for five months. The Lord was holding it for me.

> He has declared to His people the power
> of His works, in giving them the heritage of the
> nations. (Psalm 111:6)

Another amazing thing was that the duplex had been finished and for rent long before I had even prayed. Remember the scripture that the Lord gave me when He stopped the rain: "For your

Father knows the things you have need of before you ask Him" (Matthew 6:8).

Something else interesting happened when I lived there. Shortly after I moved to Baguio City the Secretary of the Party in that area was arrested. He was arrested in his apartment, which was less than 100 feet from my duplex. I believe that the Lord moved him into that place, so that he would be my neighbor. The NPA would never start trouble near that man. If it did, it might draw attention to him. God uses everything.

Then, I did something I really regret. As you have read in this account, the Lord always provided everything I needed. Rather than honoring Him by trusting Him to provide what I needed to move to Dagupan—money for deposits, rent, furniture, etc.—I called my brother in the United States and borrowed $1,000 from him.

Sometime later, after I moved in, my home church sent me a check for $1000. After I moved in, there were a lot of household items I needed, so I made a list of them. I also made a list of scriptures, including Matthew 7:7–8, Mark 11:23–24, and 1 John 5:14–15. I prayed and asked the Lord for these things based on His promises. On the list of items, I noted the date I prayed and provided a space I could note the date I received them. As I saw the need for more items, I would add them to the list and then pray and ask the Lord for them. Most days, I would pray. I would not ask for the items a second time. I would just thank the Lord that when I prayed, He gave them to me and that I believed I had received them and they were on their way. Then, I would read the list of items and scriptures to Him and myself.

For the first couple of weeks, nothing happened; then one by one, these items or the provision for them began to come in. They started to come in slowly, but as time went by, the rate increased. After a while, I lost the list. There was only one item I can remember that did not come in. It was shades for the overhead lights, but now that I think back, I had the money and could have bought them.

It wasn't long until the new class had assembled. I was able to locate a duplex near the home of one of the students to hold classes in. After we moved in, we learned that our next-door neighbor was another Bible-based cult.

Dagupan is a college town, and that cult was very active there. Cult members would go to the college campuses and invite unsuspecting students to breakfast or to visit, and when they got there, they would introduce them to their teachings. One afternoon after class, our students stayed to clean the duplex. I had given them instructions and was leaving for home. As I walked down the alley to the street, I met a young man who asked me where he could find our next-door neighbors. I didn't direct him to their side of the duplex; I directed him to our duplex and told him to ask for Manolo. The next morning, when I asked Manolo if he had come to see him, he said yes and that he had been born again. I knew Manolo would minister the Gospel to him!

A New Family

Everything was pretty routine until December. Just before Christmas, it was time to renew my visa. Since school had been dismissed for the holidays, I had two weeks with nothing to do. It had been a year since I had gone to San Carlos City and once during that time I had stayed overnight in Manila. I had enough money for my visa and round trip and hoped I had enough to stay one night in a hotel.

One of the women I had met in Manila was a travel agent, so when I got there, I thought I would call and get her recommendation on a hotel. I didn't have her telephone number, so I called the woman who had given me Etta's name to get the number from her. When I began talking to her, she invited me to come to their home. When I arrived, they invited me to stay there. They gave me a very nice room and even had servants who waited on me. But, best of all, they made me a part of their family.

The few days I stayed there were great, just like a short vacation, and there were opportunities to minister. People began giving me money and gifts. It was a wonderful Christmas, not because of the money and gifts, for they are all gone, but because the Lord was revealing His love and care for me through this new family He was giving me. He was knitting my heart together with theirs.

> So Jesus answered and said, "Assuredly, I say to you, there is no one who has left house or brothers or sisters or father or mother or wife or children or lands, for My sake and the gospel's, who shall not receive a hundredfold now in this time—houses and brothers and sisters and mothers and children and lands, with persecutions—and in the age to come, eternal life." (Mark 10:29–30)

What the Lord began on that Christmas in 1982 continues to this day.

Shortly afterward, I was invited to Manila on a regular basis to minister. I would go there by bus every other Saturday and stay until Tuesday afternoon. It was a wonderful time of ministry and fellowship.

Overcoming the Enemy

The family in whose home I stayed had a lumber lease in Mindanao, a large island in the southern Philippines. On that lease, they owned a lumber mill. The man and his wife told me that they were having serious problems with the NPA in that area. They were demanding money from them. The NPA told them that if they didn't pay, they would bum their equipment and hurt the employees. Being Christians, this couple did not want to yield to the enemies of the Lord and the government, but they knew if they didn't, they could suffer harm or loss. They were really in a dilemma.

Then, a tragedy happened that made the situation much worse. The military stationed a garrison of nine soldiers and an officer at the lumber mill to protect it. One morning, the NPA attacked the soldiers and killed all of them! It was a very grave situation. If they would attack and kill the soldiers that way, they certainly would have no qualms about carrying out their other threats.

One Saturday morning, the Lord spoke to me and said, "Tell [naming the man and his wife] if they will fast and pray, I will send the hornet." That evening at dinner, I said to them, "The Lord said if you will fast and pray, He will send the hornet."

We see hornets referred to a few times in the Old Testament:

> And I will send hornets before you, which shall drive out the Hivite, the Canaanite, and the Hittite from before you. (Exodus 23:28)

> 'Moreover the Lord your God will send the hornet among them [the enemies of Israel] until those who are left, who hide themselves from you, are destroyed. (Deuteronomy 7:20)

> I sent the hornet before you which drove them out from before you, also the two kings of the Amorites, but not with your sword or with your bow.
> I have given you a land for which you did not labor, and cities which you did not build, and you dwell in them; you eat of the vineyards and olive groves which you did not plant. (Joshua 24:12–13)

Victory Comes!

I was staying at their home each time I went to Manila to minister. On my next visit, they knew what time I would arrive, and they were waiting for me. They were almost jumping up and down; they were so excited.

They told me that right after I left, General Ramos, the commanding general of the Philippine Constabulary who was elected President in 1992, and some of his troops flew right into the airport at their lumber mill. From there, they started a "fine-tooth comb" action and swept the NPA out of those mountains!

That was in early 1983. One of the most amazing things is the completeness of this miracle. The NPA remained quite active in Mindanao. I was told that as soon as it got dark, the people in the countryside would turn their lights off and stay in their houses because they were so fearful. But in the area around the lumber mill and lumber lease, there has been no trouble at all. When I asked in mid-1989 how things were, I was told that everything was so quiet that they were thinking of starting a bank in that area! They said that every once in a while up in the high mountains, men cutting trees would catch sight of someone they didn't know who might be NPA, but no one knows for sure who they are. I believe the "hornet" is still there.

> I called on the Lord in distress; The Lord answered me and set me in a broad place.
> The Lord is on my side; I will not fear. What can man do to me? (Psalm 118:5–6)
> "Through God we will do valiantly, for it is He who shall tread down our enemies: (Psalm 108:13).

Miracles in Manila

During the spring of 1983, another family in Manila asked for meetings to be held in their home. They wanted to hold the meetings every other Sunday. Before they were born again, they had built a discotheque behind their house for their children, but when they were saved, they no longer used it for that purpose; they now used it for ministry. I would arrive on Saturday afternoon and leave for Dagupan on Monday morning.

Some people I knew in the United States had two friends who wanted to come to the Philippines to minister, so I agreed they could come and spend their time with me. These visitors were to arrive in Manila on the Monday evening after one of these ministry weekends.

The plan was for me to eat breakfast with this family and afterward go to the home of the family with the lumber lease to meet a group of women who had invited me to go on a picnic with them. (I believe that these women are some of the mothers and sisters Jesus promised to give me when I obeyed Him and left my own mother to serve Him.)

When they learned I would be in Manila all day with nothing to do, they asked if I would like to go to a beautiful volcanic lake south of Manila where one of the families had a cottage.

Everything sounded great, but I had a problem. I didn't have much money! It was the end of the month, so the monthly check could be written at any time. but I needed the money then. I only had enough money to get to the airport that night to meet the visitors and take a taxi to a hotel, but that was about it. The men were coming to visit me, so I felt I should pay their expenses and provide for them. Not only did I have to provide hotel accommodations for them that night, since I had set up ministry for them in Manila, we would be there for some days and nights. There would also be their food and transportation costs.

This was the first opportunity I had to believe God for provision for someone else besides myself, my ministry, and the students. Usually, the Lord provided what I needed in advance, but every once in a while, I had an opportunity to look to Him and wait. I would like to be able to say that in all of these situations, I stood strongly and confidently, but that was not always true.

There was the time I needed about 8,000 pesos when I got back to Baguio City after ministering in Manila. I had to pay rent, utilities, buy food, etc., when I got there. I was at a morning Bible study, and when it was over, I would eat lunch and then leave for Baguio. All I had was a few hundred pesos.

As the Bible study progressed, I became more and more concerned about where I was going to get the money. By that time, I had been in the ministry for seven years and the Lord had never failed to fulfill His Word nor was He ever late. But I was very concerned.

After the Bible study ended, I was standing by the door saying goodbye to some of the people. As I stood there, a lady came up to me and gave me an envelope. When I got it in my hand, I knew what it was, and from the thickness, I knew about how much it was. I knew it was all the money I needed. On the outside, I was calm as I thanked her, but on the inside, I was jumping up and down rejoicing. As she walked away, the Lord spoke to me and said, "You should have been doing that before you received the money."

But there were times, by His grace, that I did stand by faith and see the salvation of the Lord.

There was the time in Baguio that I had no cash at all. I had planned to go to some schools the next day and distribute tracts, but I would need transportation and lunch money. I calculated how much money I would need, and it came to about 50 pesos. Before I went to sleep, I prayed and asked the Lord for 50 pesos for my next day's expenses.

"'Therefore I say to you, whatever things you ask when you pray, believe that you receive them, and you will have them" (Mark 11:24).

I got up at my regular time, took my shower, had my normal devotion time, and got ready to go. When I was ready to walk out the door, I sat down on the edge of the bed and spoke to the Lord. I told Him I was ready to go, so I was ready for my money. As I sat there waiting, it came to me to go look in one of my suits. I went to the closet, and when I reached into the first pocket, I felt money. It was a small number of bills folded in half. When I counted them, there was exactly 50 pesos.

"For your Father knows the things you have need of before you ask Him" (Matthew 6:8). "It shall come to pass that before they call, I will answer" (Isaiah 65:24a).

Again, the Lord had foreseen the need and provided.

There was the time I was scheduled to go to Manila, but didn't have the money. Even though I was spending that weekend at the hotel and my room and food would be provided without charge, I would have to pay my transportation and tips for service while I was there.

On Friday afternoon, there was a Bible study scheduled in Baguio, so I stayed home that day. I had enough money to pay my taxi fare to and from the Bible study, but that was about all. I needed the money for the bus fare that day so I could get my advance reservation for the next morning. Even though I didn't have the money, I had great peace and the assurance that I would have what I needed.

After lunch, I went to get ready for the Bible study. The linen cabinet was beside the door going into the bathroom. In that cabinet was a small drawer where I kept my hair dryer, other miscellaneous things, and envelopes in which I kept money.

These envelopes were marked for various things: Tithe money, extra money that I didn't want to carry, rent money, money for utilities, and so forth. Since I was in that drawer at least once each day, I knew what was in it. I was the one who put the money in the envelopes and took it out so I knew what was in them.

As I stepped out of the bathroom after my shower, I opened that drawer to get out my hair dryer. When I looked into the drawer, I saw a 500-peso bill lying there. I tried to figure out where the money came from. I thought maybe it had been there the whole time or maybe it had been in one of the envelopes, but the more I thought about it, the more I realized it hadn't been there before. I bought my ticket after the Bible study, and the next morning, I left for Manila.

I had decided to spend Sunday at a shopping area passing out tracts. My plan was to go to church in the morning and from there to take a jeepney to the shopping area and spend the rest of the day. When I left the hotel, I took a taxi to church, arriving a little early. As I sat there waiting for the service to begin, I thought I would prepare my offering. I took an envelope from the back of the pew in front of me, and as I reached into my pocket, I pulled out a 5-peso bill and a 50-peso bill. That's all I had left. I guess in the back of my mind, I was still thinking that the 500-peso bill had been in the drawer all of the time and that I had somehow overlooked it. I took the 5-peso bill, put it into the envelope, sealed it, and laid it on the seat beside me. Just after I did that, the Lord spoke to me and said, "Tithe."

Then, I knew for a fact that the Lord had put the money in the drawer, because as soon as I received anything, I took the tithe out and put it aside. I always give the first fruits to the Lord. Any money that was already in that drawer would have been tithed on.

I opened the corner of the envelope, took out the 5-peso bill and inserted the 50-peso bill. The 5 pesos I had left were enough for my jeepney fare to the shopping area and then back to the hotel.

That evening at the hotel, some people gave me over 20,000 pesos. When I got the money, I took out the tithe, took out what I needed, and then prayed and asked the Lord what He wanted me to do with the rest. Shortly thereafter, He showed me the plans He had for it. (This is the money that the Lord told me to use to buy clothes for Remy and her children.) What a wonderful, wonderful life! A few times I've told people, "The Christian life is so great that even if we didn't go to heaven, it would be worth it." There are many temptations, trials, and persecutions, but knowing and living with God are worth it all. They and the treasures of this world are really nothing compared to Him.

I had always believed God could and was willing to give us what we need in that way, if it's the way He chooses. Remember Peter and the tax money? In Matthew 17:27, Jesus told Peter, "Nevertheless, lest we offend them, go to the sea, cast in a hook, and take the fish that comes up first. And when you have opened its mouth, you will find a piece of money; take that and give it to them for Me and you." Jesus knows where all of the lost money and treasures are. Think of all that has been lost in storms, fires, and so forth. I believe He can bring it to us or us to it, if it's His will.

Let's go back to the two visitors who would be arriving from the United States that night. Everyone knew they were coming and that we were staying in Manila, because I was setting up ministry for them.

As I sat at breakfast with the family, the man said to me, "I'll make hotel reservations for you. Call me later today." I thought, *What does that mean? Does that mean he pays, or does it mean we go to the hotel, and when we get there, I have to pay? Do I wait until we check out to find out about payment?* And then I thought, *Besides that, we need food*. Then, another thought really hit me: *This is a wealthy man. He's going to make reservations at an expensive hotel, and this is going to cost me a lot of money. Without him, we could have gone to a cheap hotel.* As you can see, the loins of my mind were not girded up, and I was not standing fast in faith and patience. *Now*, I thought, *I really have a problem!*

I thanked them and went to meet the women and go on our picnic. It was wonderful. I still remember the food. What a blessing! And, of course, the fellowship was great. But every once in a while, I would think about my situation. How, at times, the Lord must sit in heaven and shake His head as He observes us!

You would have to meet the Filipino people to know how hospitable and gracious they are. Once, Tom Jr. told one of the women he would like to see Manila sometime. The next morning, she and her husband came to the house where we were staying to take him out and show him Manila. Her husband had even taken the day off work. We found that you had to be very careful what you said there!

I had said nothing to my host or anyone else about my need. I knew I should take everything to the Lord and wait for Him.

After we got back from the picnic, I called my friend, and he said, "I have reservations for you at AIT [Asian Institute of Tourism]. Just sign for your rooms and food, and I will send someone over later to pay for them."

The AIT is a beautiful hotel. When the plane arrived, I found that only one of the men came. When we got to the hotel, we found three rooms were reserved for us. The Lord provides what we need, but it's up to us to use it.

When we got up the next morning and looked at the newspaper, we saw that the exchange rate on the dollar had gone from about 9 pesos per dollar to about 14 pesos per dollar, about a 55 % increase. That's by far the biggest increase I have ever seen.

Because of that, the dollars of the man who came to visit went a lot further, and so did mine. For me, this was the perfect time for the increase since I would need extra money for this man's visit. The man who didn't come missed out because everything was prepared.

Strife Enters

The man stayed for about two weeks, and by the time he left, strife had started to enter in. The day before he was to leave, I took him to Manila on the morning bus. When we got there, I had a big hassle with a taxicab driver, which is unusual for me. Because of God's grace, up until that time I never had problems with taxi drivers. When we got into the cab, I told the driver where to take us, but he wouldn't turn on the meter. He just drove away and said the fare would be 80 pesos, which I knew was at least twice as much as it should be. I told him to turn on the meter, but he refused and just kept driving. After a few attempts to get him to turn on the meter and him refusing, I told him very firmly to stop the cab. When he stopped, we got out, and I gave him what I thought we owed him, and we caught another cab.

I heard a man say once that we should avoid strife like the plague, because it will kill you. He is right.

Once, there was a lot of strife in my home that lasted for a period of two or three months. During this time, my finances really suffered, and I had to call my brother and get money from him. I believe the reason the Lord told me to name my ministry "Faith By Love Fellowship" is so that I'll always remember it is by love that faith works.

I decided the second class should be for one year. I felt this longer period would be better for the students. As it worked out, it was better, because I would be spending four class days a month in Manila, and this class would have the long Christmas and New Year's vacation in it. This class ended with just about the same amount of classroom time as the first.

When it was time to start the outside ministry for this class, Digna came back to help. She took the students out each day and conducted that part of the training. She also brought Fe, her sister, with her to go to school. Since I was going out twice a month on four-day trips, Digna and Fe could take care of Chewie and prepare Larry's meals. The Lord had put everything together.

After that class graduated, we had one more class that lasted for about eight months.

Then, we had a very sad situation. Fernando was a young man who attended the first class a couple of times, but did not continue. However, he did attend the second class and graduated. He loved the Lord and was a good student and a good minister. In the provincial areas of the Philippines, a lot of the care given to patients in the hospitals is by family members or friends who stay in the room with the patient. These people are called watchers. Fernando would go to the hospital and find patients with no watchers and stay with them and help them. That's the kind of young man he was.

Sometime during the third class, he fell into error. After he graduated, I did not see him very often but only heard of him from time to time. After a while, we heard he was going door to door and telling people that salvation was finished: That there was no more salvation. He was saying things similar to Hymenaeus and Philetus in 2 Timothy 2:16–18:

> But shun profane and idle babblings, for they will increase to more ungodliness.
> And their message will spread like cancer.
> Hymenaeus and Philetus are of this sort, who have strayed concerning the truth, saying that the resurrection is already past; and they overthrow the faith of some.

Others and I went to see him, to minister to him, but he would not listen. As I talked to him, he was correct about everything he said except that one point, and when you talked to him about that, he would just sit there and smile at you. Soon after I saw him, he died. The doctors said that his ailment was anemia and that was why he was confused, but that was not it. He had perfect understanding about everything else, except that one thing.

New Friends

I was invited to a new fellowship in Manila one Sunday morning. The pastor was the director of counseling at a well-known international Christian television and telephone counseling ministry.

When a certain woman called for counseling, she also asked if they knew of a good church where she and her family could attend, and they recommended this pastor's fellowship.

The Sunday I attended was the morning the woman brought her unsaved husband. After the man was born again that day, we met, and the Lord gave us a very special relationship. At that time, he was the resident manager of the luxury hotel I have referred to.

Getting Rid of Idols

Later, the hotel manager was born again, and after that, the resident manager's two sons were born again. One night when I was visiting them, we were talking about how demonic some comic books are and that Christians should not read or keep that type of comic book.

The oldest son, about fifteen, said that he had a large collection of comic books, and some of them were quite valuable. When he realized they were not pleasing to the Lord, and he should not sell them, he said that he wanted to burn them.

We spent more than an hour that evening burning comic books. His little brother stood there the whole time in amazement, not believing that all those great comic books were being burned up, but this young man was glorifying God.

We see in Acts 19:19 what the Ephesians did: "Also, many of those who had practiced magic brought their books together and burned them in the sight of all. And they counted up the value of them, and it totaled fifty thousand pieces of silver." It's wonderful to minister to those who are that sincere and will give all for Jesus.

Another of my friends who was born again knew she would have to get rid of a certain religious idol, but it was so expensive she did not destroy it; she sold it for 30,000 pesos. Shortly after that, she had a financial loss of 30,000 pesos!

An Impossible Prayer

One evening in Dagupan, as I was reading the Bible, I came to Romans 15:20 that says, "And so I have made it my aim to preach the gospel, not where Christ was named, lest I should build on another man's foundation."

I said to the Lord that I also desired to preach Christ where He has not been named. Then, I thought, *Where could I do that? Maybe on an island somewhere, but how could I speak the language?* The more I thought about it, the more impossible it seemed. (I should have quit thinking about it and simply trusted the Lord.)

Years later, I was invited to minister to a group of people from another country. They had been sent to the Philippines for training.

As I ministered to them, many were born again. I remember being told by some of them that they knew about the book Tom Sawyer, because they had read it in their schools. They said that they knew all about our first president, George Washington. They said, "He was the father of your country." And they said that they knew about Abraham Lincoln. They said, "He set the slaves free. He was a great man." But some of them had never heard the name of Jesus! The Gospel of Jesus Christ had been repressed in their country.

These were college-educated business professionals who had traveled in their country. They were not uneducated people from remote areas. They knew all about a fictional character from another country, but they had never heard about the God of the universe. They had heard about a father of the United States, but they had not heard of the Father of all mankind. They had heard about the man who had set the slaves free in the United States, but they had never heard about the Man from heaven who had set all of mankind free!

It was then I realized what a great conspiracy there is in this world, "Against the Lord and against His Anointed" (Psalm 2:2), against our heavenly Father and the Lord Jesus Christ.

This same conspiracy has been here since the beginning.

> But so that it spreads no further among the people, let us severely threaten them, that from now on they speak to no man in this name.
> And they called them and commanded them not to speak at all nor teach in the name of Jesus. (Acts 4:17–18)

But this conspiracy will not prevail.

> And this gospel of the kingdom will be preached in all the world as a witness to all the nations, and then the end will come. (Matthew 24:14)

The Answer Comes

Once, the man in charge of this group told them that they could not come to our meetings, Bible studies, anymore. He told them their country was not Christian but instead named another religion, and if anyone continued to come to the meetings, it would be put in his or her record. When I returned the following week for our Bible study, this man had been relieved of his duties and sent home. The man who was put in his place permitted the visitors to continue to come to the meetings.

Something happened that I thought was very interesting. After we ministered to the first group, they sent a second group. Right after they arrived and before we had an opportunity to minister to them, a Christian minister of their nationality was asked to speak to them. Two from our ministry were also invited. He spent a considerable amount of time speaking to them about the Lord and how good He is, inviting them to "taste and see that the Lord is good"; but when he came to the point where a minister would normally give an invitation, he quit speaking. After he sat down, the person who was with me was asked to speak to them, and she invited them to receive Jesus. All but one received the Lord.

What I thought was so interesting was that the minister did not pray with them, but we were given that opportunity. I am convinced the Lord led me to pray that prayer years before, and He had reserved the opportunity for us to pray with them.

This is borne out in scripture. In Deuteronomy 2:9, the Lord told Moses, "Then the Lord said to me, 'Do not harass Moab, nor contend with them in battle, for I will not give you any of their land as a possession, because I have given Ar to the descendants of Lot as a possession.'"

In the book of Numbers, chapter 22, we see the Israelites camping in the plains of Moab. But Balak, the king of the Moabites, did not know that the Lord had given the land to them (the Moabites) as a possession and hired Balaam to come and curse the Israelites for him. Had he known God and what He had promised them, he would have received Moses and the Israelites as who they were, people sent by God. The Israelites had defeated Sihon, the king of the Amorites, who previously had taken some of the Moabites's land. Had he just waited, Moses and the Israelites would have moved on, and then, they could have returned and occupied the land God had given them.

I believe that what God has given to us, He will preserve for us and will not give it to anyone else. And if we are faithful stewards, this includes our ministries.

We baptized some of these new Christians in a swimming pool, and when we could no longer do that, we baptized them in bathtubs!

A New Convert's Prayer

When the third group came, because they went straight into their training schedules, we did not have an opportunity to hold an evangelistic service for them. All we could do was hold a Sunday evening Bible study for them. One Sunday evening, one of the women came to the Bible study for the first time. Right after she arrived, someone shared the Gospel with her, and she was born again.

At the end of the Bible study, each one was asked to pray a short prayer. When this woman's turn came, she said softly and reverently, "Father, I ask You to forgive me, but I didn't know You were there." And to think that there are multitudes in this world just like her!

> Then Jesus said to His disciples, "The harvest truly is plentiful, but the laborers are few.
> Therefore pray the Lord of the harvest to send out laborers into His harvest." (Matthew 9:37–38)

While we are on this point about laborers, I would like to share something with you. Jesus said to us in John 4:35–36:

> Do you not say, "There are still four months and then comes the harvest"? Behold, I say to you, lift up your eyes and look at the fields, for they are already white for harvest!
> And he who reaps receives wages, and gathers fruit for eternal life, that both he who sows and he who reaps may rejoice together.

Here, He clearly says that God intends to pay wages to His laborers. As we labor in His fields, we can expect to receive our wages. He has made us this promise.

Not only has He promised to pay us wages, but also He is unhappy with those who defraud their laborers of their wages. James 5:4 speaks to those who defraud their laborers:

> Indeed the wages of the laborers who mowed your fields, which you kept back by fraud, cry out; and the cries of the reapers have reached the ears of the Lord of Sabaoth.

The Lord also said in Malachi 3:5, "'And I will come near you for judgment; I will be a swift witness Against sorcerers, Against adulterers, Against perjurers, Against those who exploit wage earners and widows and orphans, And against those who turn away an alien—Because they do not fear Me,' Says the Lord of hosts." God will never defraud His laborers.

Over the years, I was invited to spend many of my visits to Manila at the hotel. It was always a wonderful time of refreshing, and that family and hotel blessed me often and in many ways. The Lord also used them to provide ministry opportunities. I will tell you of some of them.

The Testimony of Jimmy Go

When I first got to San Carlos City in December 1981, the man who cashed my checks gave me a book to read. The title of the book was *Jimmy Go*. Mr. Jimmy Go was the editor of the Fukien Times newspaper, which was published in Manila prior to World War II. Because he was able to discern Japanese intentions for Asia before the war broke out, Mr. Go wrote many articles in his newspaper exposing them. Because of this, he was high on their list of people to be dealt with when they invaded the Philippines, and he knew it.

When the Japanese invaded, his wife was a committed Christian, but Mr. Go had not yet fully committed his life to Christ. The book details how Mr. Go fully committed his life to the Lord and how He led and protected his family and himself throughout the war years as the Japanese continued to search for them. It was a great blessing for me to read that book. As Mr. Go, I was also totally dependent on the Lord. I was very encouraged by it.

About six years later, through the hotel manager, I met one of Mr. Go's daughters. She was the editor of a number of newspapers in the Philippines. Later, I met Mr. and Mrs. Go before he died and had opportunities to fellowship and pray with them.

At my first meeting with Mr. Go's daughter, I made arrangements to print my tract in one of her newspapers. That time, I paid for the printing, but she told me later that from time to time, they print the tract in the newspaper on their own. I have no idea how many people have been reached with the Gospel through this method. Those papers are distributed throughout the Philippines, and some are sent to other countries.

When the Lord gave me the tract ministry in 1982, He told me not to put any name on it. He said that He wanted it to be a ministry directly from Him to each person. He did not want any person's or any organization's name to interfere with anyone reading it and believing what it said and receiving Jesus.

When I started to print the tract, I used a printing company near Dagupan. That first company did a good job, but when I met Mr. and Mrs. Go's daughter, Mrs. Betty Go Belmonte, I made arrangements with her to print the tracts. It was a much better arrangement since I rarely went to Dagupan anymore and I was in Manila every weekend. Also, I was dealing with a Christian publisher.

When I made arrangements with her to print the tract, she gave me a price of five centavos each, which was a much better price than I could get anywhere else. By then I was printing the tract in many different languages, so this lower price helped me to keep up with the ever-increasing demand.

She and some of her employees began to come to the Bible studies that were started at the hotel. One day, she told me that from then on, they would print the tracts for one centavo each. That meant a hundred dollars in American money would buy about 260,000 tracts! As my needs increased, so did the Lord's provision and always in ways I never would have thought of.

CHAPTER 6

A New Direction

When I was on my way to the Philippines, I met a couple in Phoenix, Arizona, and told them if they ever wanted to come to the Philippines, they could stay with me. They contacted me and said that they would like to come over and minister during February and March of 1984. I agreed, and in the sixty days, they were there, they ministered about eighty times.

A side note is that they had been married for many years, but were unable to have children. While they were in the Philippines, she became pregnant and had a daughter.

While they were there, a woman in Manila contacted me and told me that she would like to send me to Baguio City for a vacation. Since the people who were visiting were well able to teach school, I felt free to go.

She made arrangements for me to stay at the apartel of one of her relatives. (An apartel is a hotel of fully furnished and equipped apartments for daily, weekly, or monthly rental.) She also gave me money so that I would have an enjoyable time. I didn't see the Lord in all of this at the time, but I later realized He had set everything up. He was making preparations to move me to Baguio City in fulfillment of what He had told me years earlier.

Demonstration of Power

I had a wonderful time. One day, I took a taxicab to Asin Valley to visit the wood carver shops. After I had browsed through the shops, I started to walk up the road out of the valley. As I walked, I came to a small boy, about eight years old, leaning against a guard rail, reading a comic book. A colored comic book in that area is a very valuable possession.

When I got to where he was, I gave him a tract. He looked at the tract for a few seconds and then threw the comic book down and started reading the tract. I was amazed when he threw the comic book down, knowing how important it probably was to him. I did notice that it had a tear in it, but other than that, it was almost brand new. When I saw that happen, I began to more fully realize the power of God's anointing when we minister.

The owner of the house from whom I rented the upstairs in San Carlos City was a retired school principal. His wife was a first-grade teacher who had been teaching for thirty years. She was rated at the top of their rating system of qualifications for Philippine schoolteachers. She was a very good teacher.

In the Philippines, because the weather is warm year around, the preschool children spend most of their time outside playing. They don't spend very much time sitting quietly. Because of this, when they enter school, they are very active.

In this woman's classroom, there were two teacher aides. These young women helped to get the children quiet and attentive so the teacher could teach them.

One day, she told me about the time two of our students came into her classroom to minister. You can tell she did not recognize them from what she said. She said, "I was amazed. Two young men came into my classroom to talk to the students. When they came in, I went and sat at the back of the class. It was amazing, because while they were there, the children were all quiet. I could not see where their eyes were focused, since I was behind them, but all of their heads were directed at the young men." They were in that classroom a little over a month after school began. That is the power of God.

Another example of this was when I first got to the Philippines. One Sunday morning, I was invited to a church to preach. I repeatedly noticed one man who kept falling asleep. All through the service, he would fall asleep for a while and then wake up. Then, his head would fall over and he would be asleep again. Of all the people who were there that morning, he was the last one I would have thought would hear or receive anything, but when I gave the invitation, he came forward and received Jesus.

A similar situation occurred several years earlier when we were showing the Christian movies in Tulsa. One night, we were showing a foreign film, whose visual quality was so dim, spotted, and streaked that it was hard to see. Also, with the heavy accents, it was difficult to understand. Between 400 and 500 people were there that night. As the movie continued, people began to get up and walk out. As I watched the people leaving, I became discouraged. The main reason for showing the movies was for people to be born again. If they left, they wouldn't be there to receive the Lord at the end of the movie. By the time the movie ended, at least half the people had left. But when I gave the invitation, three times as many came forward to receive Jesus as had ever came forward before or after that evening.

I learned something through these situations. Don't be moved by what you see. Also, I clearly saw that the Lord is in control and is the One who saves.

Once when I was in Manila, I heard a message given by a well-known international evangelist. The message he preached was so simple that it seemed to me any Bible school student could preach it, but the anointing that was on that Word was powerful. That night, I saw 1 Corinthians 2:4 in demonstration:

> And my speech and my preaching were not with persuasive words of human wisdom, but in demonstration of the Spirit and of power.

It is God and His anointing that does the work.

Going Forward

While we're on the subject of going forward in a service to receive Jesus, I would like to insert something here. Going forward has nothing to do with salvation.

The scripture simply says:

> But what does it say? "The word is near you, in your mouth and in your heart" (that is, the word of faith which we preach):
> that if you confess with your mouth the Lord Jesus and believe in your heart that God has raised Him from the dead, you will be saved.
> For with the heart one believes unto righteousness, and with the mouth confession is made unto salvation.
> "Whoever believes on Him will not be put to shame." (Romans 10:8–11)

> For by grace you have saved through faith, and that not of yourselves; it is the gift of God, not of works, lest anyone should boast. (Ephesians 2:8–9)

There is nothing wrong with having people come forward to receive prayer, but we should never make it a requirement for salvation.

Once while on a bus in the Philippines, I met an elderly man who, as we talked, invited me to his home for a Sunday lunch. When I accepted his invitation, he offered to pick me up in his car and take me to his home, which was in another part of our province. I told him I would be in church until about noon, but would be free to go with him after that. When I told him this, he said that he would pick me up at the church.

He and his driver arrived early, shortly after the service started. When they came to the church door, they told whoever was at the door that they were there to see me. I was sitting at the back so the persons they met at the door brought them in and seated them in some empty seats in front of me. The pastor was preaching an evangelistic message that morning. When he was finished, he asked everyone to stand up. As we all stood there he said that if anyone wanted to receive Jesus to put up their hand. The two men who came to pick me up both raised one of their hands to a little above their waists. Then, the pastor invited those who had raised their hands to come forward to receive Jesus. They and some others that I had noticed raising their hands did not go forward. Since they did not go forward, they did not receive salvation.

After the church service was over, they took me to his home, and we had an enjoyable meal and visit. When they drove me home, I asked them if they would like to come in for some desert, to which they consented. As we sat there talking, I told them I had seen them raise their hands but knew they had not prayed. I then asked them if they would like to receive Jesus and they said yes. After that, we prayed and they were born again.

After my vacation in Baguio City was over, the Lord clearly showed me it was time to move there. Actually, He said to "go to the mountains," but I knew it was Baguio City He was referring to.

I began to ask some of the people I knew if they knew of any housing for rent in Baguio. One lady told me that she had a house near the market that she would rent to me, but when I tried to go there with her to look at it, something came up, and she was not able to go. She never was able to go, so I finally gave up on it.

Two Special Blessings

Then another lady told me about a brand-new house her cousin was building. She would be renting part of it. When I went to Baguio to look at that house, they were just beginning construction. The people who were building the house were two Filipino women who had worked in Chicago, Illinois, for many years. One woman was a nurse who was still working there. The other, who was constructing the house, was a dental technician who had retired and moved back to the Philippines.

She is quite a lady. When I got there, she had a group of men working. She totally constructed that house, buying the materials, hiring the labor, and supervising the construction. Having been in construction myself, I know what a major feat it was. She built a beautiful big house.

I rented an apartment from these women for almost six years, and they were outstanding landladies. They were very kind and attentive, like two elderly aunts who couldn't do enough for me, two of God's special blessings.

Actually, all of my landlords have been great, but I guess the reason I was so impressed with these women was because they were ladies and because of their age. Being in the professions they were in and having lived in the United States for so many years, they knew what Americans like and how to make us comfortable. How good God is!

The house was being built on the side of a mountain, about half a mile from downtown, a quiet spot with a beautiful view. When you entered the property from the street, you entered at the upstairs level. This floor would be their home. Then, they built concrete steps down on one side of the house to the downstairs. They planned to build two apartments downstairs that they would rent out.

When Dada was beginning the interior construction downstairs, I asked if she could leave some of the walls out and make one big apartment. I told her if she could do this, I would rent both apartments. She agreed. When the construction was finished, I had a very nice, large apartment.

Where one apartment would have been, I had the kitchen/dining area, living room, and a bathroom. I used the area that would have been two bedrooms for the living room. Where the other apartment would have been, I had a small office built and used the rest for a bedroom. In one corner, I had a combination study with built-in bookshelves and dressing area.

I also had a built-in linen cabinet and built-in dresser. There were two closets for clothes and a full bath. A rattan bookcase was converted into a nice bookcase headboard for the bed. Later, I had a carpenter come and build a trunk that sat at the foot of the bed. I brought cedar from the United States on the second trip so it was set up like a cedar trunk. I also had him build a buffet for the dining area.

A few days after I moved in, the Lord gave me about 15,000 pesos and told me to go to Dau and buy some rattan living room furniture.

Dau was a shopping area near Clark Air Force Base that had rattan manufacturing and sales stores that sold beautiful export-quality rattan furniture.

I took a bus from Baguio to Dau and went into one of the stores that had a lot of furniture groupings displayed in a large showroom. After finding the right one, I placed my order and then was sent to select the fabric and the colors for the pillows. It was a lot of fun. What a blessing to serve the Lord!

When the furniture arrived, some of the living room furniture I had brought from Dagupan was made into a sitting area in the bedroom. A concrete deck ran the full length of the south side of the apartment and halfway down the west. It was wonderful to be able to sit there and enjoy the beautiful scenery, sunshine, and weather. The Lord really gave me a nice place. And again, screens were put on the windows free of charge, even though I rarely saw an insect that high up in the mountains.

Writing about the rattan furniture reminds me of something else the Lord did. As I was preparing to leave on the second trip to the Philippines, I went shopping. A Thom McAn shoe store I went into was displaying a new line of men's shoes. This was a new type of shoe with a molded and flexible sole with heel combined, and the upper was a very nice soft leather. The pair I saw that I liked were light tan loafers, but when I saw that the price was $75, that was it. I didn't want to spend that much money for a pair of shoes right then.

After returning to the Philippines, I spent more and more time in Manila, and much of it was spent at the hotel. It became obvious I needed some new clothes. The hotel is a five-star hotel and is considered to be one of the finest in the world. I was coming into contact with well-dressed people, and I needed to dress appropriately.

As the Lord provided, I bought new clothes. One day, when I was at the hotel, someone gave me 2,000 pesos, so I decided it was time for me to buy new shoes. After deducting the tithe, I took all the money, even though I never intended to spend that much for shoes, and went to a large shopping center near the hotel. When I got there, I found a shoe store and went in and began looking around. All I remember seeing were Filipino- and Italian-made shoes. As I was looking, I saw a pair of the Thom McAn loafers I had wanted to buy in Tulsa. Not only were they the same shoe, but also they were the same color, and when I tried them on, they were the right size. Then, I thought I would check to see if they might have a pair in another color. I asked the salesman and he said, "No, this is the only pair they sent over." The price was 1,500 pesos, the same price the pair in the states had been. I never saw another pair of Thom McAn shoes in the Philippines in any style, color, or size.

When Jesus sent the twelve out in Matthew 10:8–10, He said, "'Heal the sick, cleanse the lepers, raise the dead, cast out demons. Freely you have received, freely give. Provide neither gold nor silver nor copper in your money belts, nor bag for your journey, nor two tunics, nor sandals, nor staffs; for a worker is worthy of his food." I don't wear sandals, but I do wear loafers.

There Is Shooting Ahead

It was early May when I made arrangements to rent the apartment in Baguio City. The moving date, June 24, was about two months away, and Dada said that she thought my apartment should be ready by then. Since school was closed, I spent that time in Dagupan in prayer and Bible reading or in Manila ministering.

Once while I was in Dagupan, I had to go to Baguio to see Dada about something. Tom was there and offered to drive me. We took the new Marcos Highway that had been built as a scenic drive to Baguio. It is beautiful. In some places, the highway is cut out of the sides of the mountains. As you wind up toward Baguio, you can look back and see that flat countryside covered with a patchwork of forests and fields of various colors and then behind that you can see Lingayen Gulf spread out before you.

It is a lonely road and seldom used. We knew there were NPA all through that area, but by God's grace, we had learned to live and travel without fear. When we were well up into the mountains, we came around a curve, and about 500 or 600 feet ahead of us, we saw a large group of men with weapons and an armored personnel carrier. They were irregularly dressed—some were wearing military fatigues, and some were wearing civilian clothes. They definitely were not Philippine military. As soon as we saw them, Tom asked, "Should we turn around and go back?"

Before I even had time to think about it, these words came out of my spirit:

> A thousand may fall at our side, and ten thousand at our right hand, but it will not come near us. (Psalm 91:7)

As we pulled up to the men, they motioned for us to stop. When we stopped, one of them said to us, "Will you wait here for a while? There is shooting ahead."

A New Direction

As we sat there, we began talking to them. They seemed to be pretty nice guys. We also gave them tracts. After about fifteen minutes, they told us we could go ahead. As we drove on, we came around another curve, and there was another group of people dressed the same as the first. They had movie cameras and were making a movie. When you are out in a situation like that and someone tells you there's shooting ahead, the last thing you think about is shooting a movie! As we went by, they smiled, waved, and thanked us. Again, the Lord had given me utterance, and through that Word, He had given us the courage and strength to move on.

The other day, I saw a bumper sticker that read, "Sky diving is pure adrenaline." I'll tell you something that is better than that: Walking with God by faith.

Once when I was in Baguio, a man walked past me carrying a little boy in his arms. The sidewalks were full of people, and the street was full of traffic. It was bustling and noisy, but that little boy just lay there in his daddy's arms with his head on his shoulder, sleeping. When I saw that, I believe the Lord was saying to me, "That is the way you are to be." Remember Jesus in the boat.

Living with and walking with the Lord is the greatest, most exciting life there is, because He is great and exciting. Our Lord Jesus said that we are in the world, but we are not of the world. We live in a natural world, but the Lord has given us a supernatural life.

Attacks by the Devil

As it got closer to the time to move, the Lord spoke to me and said that when I got to Baguio, I would have to move into temporary housing. Dada said that although she was having some problems, she was still trying to get the apartment ready for me to move into on schedule. The devil was also working, trying to hinder. Strife began to break out everywhere.

Then my passport was stolen! Etta, the woman at the Immigration Office, had gotten my visa extended and was bringing it to me. On the way home from work, she went to a church service. While she was there, someone broke into her car and stole a number of things, including my passport. American passports are valuable, and some people sell them. Because of that, when I went to the American Embassy to get it replaced, they treated me like a criminal.

I didn't really blame them for being suspicious, because of what some people do; however, Etta was totally innocent. Her being involved was the worst part of the whole thing. She had always been such a blessing and helped me so much, and now, she was caught in the middle. That's what happens when you get into strife. It not only hurts us but also can hurt others around us. Strife is one of the most devastating sins there is.

They arranged a hearing for Etta and me. The man who held the hearing was very rough on Etta. If I had walked in love and kept the door closed to the enemy, I know in my heart my passport would not have been stolen.

However, the Lord is merciful and continued to supply my needs. I would need an additional 1,500 pesos a month for rent, a deposit on the apartment, moving expenses, new things I would need for the apartment, and unexpected expenses, but I didn't give it much thought.

After I had moved to Baguio, I went by the house one day, and Dada told me she had been told by some people there that she should have me pay an additional two months' rent in advance. They were right—three months' rent in advance is a standard practice there. Because neither Dada nor I had rented property there before, we were unaware of it. Since the house was part Tessie's (Tessie is the nurse who stayed in Chicago working), I'm sure Dada felt that she should follow all of the proper procedures.

By God's grace, I had learned to trust Him for my needs, and I no longer was concerned about them. One day when I was in Manila, a man and his wife came by the house where I was staying and asked for counseling. We had an enjoyable visit, and when they left, the man gave me an envelope. It was obvious from the feel that it was money. Generally, in the Philippines, when someone gives me a gift, it will be in 100-peso bills. From the thickness of the envelope, I thought it was probably about 5,000 pesos. However, when I took the money back to my room and counted it, it was over 14,000 pesos. Most were brand-new bills. At that time, 14,000 pesos was worth about 1,000 American dollars. With that money, I had all that I needed to move. I used none of that money to buy the rattan furniture. The money for the furniture came later.

Some friends volunteered to come to Dagupan with their truck and help me move. As I was thinking about the move one day, I wondered what the back of the truck was like. If it was the Philippine standard, the back of the bed would be closed in with a door built in for entry; however, some of those trucks have tailgates. If it was the door type, we would not be able to get the refrigerator in. If it were the tailgate type, we would. I had seen the truck before, but I couldn't remember what the back was like. As I sat there thinking about it, the Lord gave me a vision. He showed me the back of the truck clearly.

It was the tailgate type. We were all right.

I was not ready to move on June 24, but I was ready by the next Saturday. We left with the first load early Saturday morning.

They had major problems trying to leave Manila, and I had major problems trying to leave Dagupan. This strong resistance went on for months, and much of it was manifested as strife and contention. I had gone to Baguio on the previous Sunday to pass out tracts, and while I was there, I went by to look at the apartment. It was a long way from being ready, but I had to move by the next Sunday, so I continued to move ahead, and Dada continued to do all she could to be ready.

When we got to the apartment with the first load, I saw immediately that I could not move in. And moving back to Dagupan was out of the question. Then, I thought about the apartel. I said, "Let's go to the apartel and see what we can do." When we got there, they welcomed us with open arms. They gave me a place to store the furniture and an apartment to live in, and they would not take any money. They also said that if I wanted to, I could stay at the apartel permanently, free of charge, but I had already made arrangements at Dada's and Tessie's, where I knew the Lord wanted me to live.

Later, I remembered that the Lord had said that I would be taking temporary housing when I arrived in Baguio.

The Lord's Preparations

When I first arrived in the Philippines, I found the Lord had gone ahead and prepared the hearts of the people to receive me. The Lord put me with a group of people who truly love me, and their chief concern in our relationship is my welfare, with little thought for their own. How wonderful it is to be truly loved! I have experienced God's love there.

I thanked the people at the apartel for their hospitality and concern, but I explained to them that I had already made the other arrangements; however, I did spend some weeks there.

Right after I arrived in Baguio, a group asked me to teach a Bible study. While I was preparing, the Lord told me that they would ask me to be their regular teacher. He said to tell them I would, but not for an extended period of time. They did ask me, and I told them what the Lord had told me to say. I did not know it at the time, but in a few months, I would be leaving for the United States for a while.

Chewie and Dada were becoming very close. She thought it was cute how he would come up to her house at dinnertime and sit on the floor beside her chair and eat prawns and other expensive food that he would never get at home! He was getting fatter and fatter. One day, I jokingly told her I was going to have to put him on a diet. "But," she responded, "he enjoys it so much."

Tessie remained in the United States to work for a few more years, so Dada enjoyed Chewie's companionship. I believe the Lord was putting them together, because I needed someone for him to stay with when I was gone on trips. Dada is the one who took care of him for the year and seven months I spent in the United States.

When I moved to Baguio, because of my responsibilities there, for a few months, I could no longer go to Manila on ministry trips. I was really missing the visits with my friends, the family dinners, and so forth. One week during this time, I was running out of money. Early the next week, rent was due, and I also had other financial requirements coming up.

Then, on Friday, some of my friends from Manila stopped by. It was great. They arrived in the afternoon, and that evening we went to the Japanese restaurant at the Hyatt Hotel and had dinner. Then, we spent most of the day Saturday and Saturday evening together.

Before they left on Sunday, they came by and visited for a while and left an envelope that contained all that I needed. The way the Lord ministers to us is always alive and creative. That's one of the reasons why I enjoy Him so much. Oh, the depth of Him and His wisdom!

Another thing that the Lord did was increase what I spent for food. I had been budgeting 3000 pesos each month for food. Right after arriving in Baguio, the Lord spoke to me and said to add another 1000 pesos a month to the food budget. I told Him, "Alright," and that I would. I didn't have the extra money on hand to do this, but a couple of days later, I received a check from Larry in Dagupan for 1000 pesos. That is the only time he ever sent money to me. Every month from then on, the extra 1000 pesos was always there.

Then there was the time I was in Manila at the hotel and was scheduled to leave for Baguio early the next morning. Shortly after I arrived, rent would be due, 4,000 pesos. I had enough money to pay for transportation to Baguio and food when I arrived, but not enough to pay rent. Since I would be leaving very early in the morning, I would certainly not be seeing anyone before I left. The evening went by, and I did not receive anything from anyone. I returned to my room somewhere around ten o'clock and went to bed.

At about eleven o'clock, an envelope from the manager and his wife arrived. When I opened it, there was 4,000 pesos.

For years, I had wanted to do something for the poor at Christmas, but had never had the opportunity or the provision. One year, I prayed and asked the Lord for 1,000 pesos to minister to the poor.

A few days before Christmas, I was leaving Manila for Baguio. The regular buses sometimes left a little late, but the air-conditioned buses always left right on time. You could set your watch by them.

I was sitting in an air-conditioned bus at the terminal. When it came time to leave, the bus didn't move. We kept sitting there and sitting there. Then, all of a sudden, the woman who had sent me on vacation to Baguio came running onto the bus. She said that she had been rushing and had hoped not to miss me. She gave me some things for Christmas and a check for 1,000 pesos. What she gave me was very encouraging, because I would be spending this Christmas in Baguio and separated from my friends in Manila. There is no doubt in my mind that it was from the Lord. The Filipino people are so generous and hospitable by nature that the Lord had to do many things in an outstanding way so I would know they were from Him.

When I got to Baguio, I realized the 1,000 pesos was my money for the poor. I went to the bank and got fifty 20-peso bills. I then bought fifty envelopes and put an Ilocano tract and an English tract and a 20-peso bill in each of them.

I planned to go downtown the afternoon of Christmas Eve and distribute the envelopes to the poor. Just before I left, a pastor and his wife came to visit, so I gave them twenty envelopes to give to the poor they knew.

Before I left, I prayed and asked the Lord for His direction. They had closed the main downtown street and set up an open market, and the city was packed with people. It took a long time of walking and looking before I was able to give away the thirty envelopes. There weren't as many poor people around as I thought! I realized that day that the real need of the people in that area was not so much physical as spiritual and that the main emphasis of the ministry was to be on ministering the Gospel.

Then, the Lord did something special. In the Philippines, it is customary to give someone who is visiting you something to take home with them (usually a food item). I was frequently having visitors and did not usually know that they were coming, so I rarely had anything special to give them. I wanted to be able to give my visitors something, so I talked to the Lord about it. Shortly after that, a vegetable vine began to grow up my patio fence. I don't remember the name of it, but it is very good and popular there. It was very fruitful, and from that time on, I always had plenty to give everyone who came to visit. Dada said that she couldn't understand it. She had a vine at her house and rarely had enough for her and her two maids. Sometimes, I even gave her some when she didn't have enough.

The Lord Stopped the Rain

There was another way the Lord provided. A few times, He stopped the rain!

Once, He stopped the rain at the Christian school where I worked. One afternoon, we had part of the roof of the auditorium open, and it clouded up and started to rain. If it had continued to rain, it would have ruined the insulation and acoustical ceiling tile below, so I exercised my authority over the rain and commanded it to stop in the Name of Jesus, and it did. The second time was at the steel company that night.

The next time was at the outdoor meeting with Mrs. Cruz at Dagupan, when I first arrived in the Philippines. On the first evening, just as the meeting was starting, large raindrops began to fall, so I again used my authority in the Name of Jesus, and it stopped.

Once, during the monsoon season, as I was traveling back to Baguio, it was raining so hard that water was almost at the top of the curb.

My gate was locked from the inside. So this meant I would have to ring Dada's doorbell, and her maid would have to let me in through their gate. Even though she would have an umbrella, she would get wet because of the heavy rain. So I prayed and asked the Lord to please stop the rain so she wouldn't get wet. When the taxi stopped in front of the house, the rain stopped. I rang the bell, and the maid came to the gate and let me in. Just as we stepped under the roof overhang, going up on the porch, it began pouring again.

On another occasion, I was in La Trinidad, a town near Baguio, distributing tracts. About midmorning, I was near an apartment complex passing out tracts when it began to rain. I prayed and asked the Lord to stop the rain until I was finished, and the rain stopped. I spent the rest of the morning and part of the afternoon going to various places distributing tracts. Just as I was stepping into the jeepney to go back to Baguio, it started to rain again—literally the second the ministry was finished.

I would like to share something fun that happened on that trip to La Trinidad. After arriving in town that morning, I went and stood on the sidewalk in front of a college and distributed tracts to those who entered, and to those who passed by. As I stood there a very small, elderly Nun passed by, so I gave her a tract. About 15 minutes later she came back and asked for more, so I gave her about 200. After I left the college, I walked down to an intersection so that I could cross the main street. There were 4 lanes of solid traffic (2 each way), and they were traveling every bit of the speed limit. I waited and waited, but the traffic was so heavy, and they were traveling so fast I couldn't possibly cross the street. As I stood there wondering what I should do, all of a sudden, I felt someone take hold of my left arm. I looked over, and then down to see who had ahold of my arm. It was the Nun who I had given the tracts to. The next thing I knew she was pulling my arm, and across the street we went. We didn't stop and I don't think she even looked at the traffic. I mean, everybody stopped, except us; if they hadn't we would have been killed. When we got to the other side she let go of my arm and walked away. I didn't even have an opportunity to thank her. I guess she thought, "We're about the Lord's work, and we don't have time to wait around." What an awesome woman she is. You talk about being bold, and trusting the Lord. When I get to heaven, I want to look her up. I'd love to meet and talk to her. You meet so many wonderful people in the Body of Christ.

I saw a sign on a truck the other night that read, "Jesus is Awesome." That certainly is true.

I often stayed with a family in Manila named Mendoza. They gave me a nice room to stay in and treated me like family. They took me on family excursions and activities, to dinner, and so forth and would never let me give them any money to help pay for my expenses. Since I wanted to do something nice for them, I decided to invite them to dinner. I wanted to take them to a really good restaurant. I knew that a dinner like that would cost about a thousand pesos.

A few days before we were to go to dinner, I was in Manila staying at another home. I spent the afternoon in my room, and during that time, I made a list of my financial needs. With the tithe included, the total was about 7,800 pesos. I prayed and read the list to the Lord and asked Him to provide this amount of money.

Later that afternoon, I was given an envelope containing money, and when I counted it, the amount was a little over 8,800 pesos. As I finished counting and saw the total, the Lord spoke to me and said, "You forgot about the dinner." He was right—I had forgotten about the dinner and had left it off the list. But He provided for it anyway!

Healed of Cancer

Warren and Carol are the man and wife in whose home I taught the Bible study in Baguio. One day, I saw Carol downtown, and she told me that her cousin had come to visit them. She said that her cousin had cancer and had been given three months to live. It had started as breast cancer but had now spread to the rest of her body. Carol said that her cousin had worked at the hospital in Naga City for a number of years, and they had done all they could for her, but they had finally given up. She asked if I would come by and pray for her, and I told that her I would be by that evening.

That afternoon, I went through the New Testament and listed the scriptures about the healing ministry of Jesus. I had a strong witness in my spirit that Carol's cousin would be healed. The knowledge that she would be healed was so real that it was almost tangible!

When I went there that evening, we spent about an hour going through those scriptures. Afterward, I took her to Mark 16:17–18, where the Lord Jesus said that in His Name, we would lay hands on the sick and they would recover. I asked her if she believed that, and she said yes. So I laid my hands on her and prayed. Then, we had some refreshments, and after a short visit, I went home. All evening, there was no sense of any moving of God. We just looked to His promises, believed, and prayed.

About ten days later, I received a postcard from her with a short message. It read, "Praise Jesus, I'm totally healed." About four years later when I saw Carol, I asked her how her cousin was doing, and she said that she was doing fine.

Jesus Healed Everyone

I believe we should expect these results every time we minister to a sincere, believing child of God. In Ephesians 1:3, the Bible says, "Blessed be the God and Father of our Lord Jesus Christ, who has [past tense because He has already done it] blessed us with every spiritual blessing in the heavenly places in Christ."

> For all the promises of God in Him are Yes, and in Him Amen, to the glory of God through us. (2 Corinthians 1:20)

When Jesus ministered, He healed them all.

> Then Jesus went about all the cities and villages, teaching in their synagogues, preaching the gospel of the kingdom, and healing every sickness and every disease among the people. (Matthew 9:35)

> Then His fame went throughout all Syria; and they brought to Him all sick people who were afflicted with various diseases and torments, and those who were demon-possessed, epileptics, and paralytics; and He healed them. (Matthew 4:24)

The Lord Has not Changed

> Jesus Christ is the same yesterday, today, and forever. (Hebrews 13:8)

We also see that in the early church, after Jesus returned to heaven, all were healed.

> Also a multitude gathered from the surrounding cities to Jerusalem, bringing sick people and those who were tormented by unclean spirits, and they were all healed. (Acts 5:16)

We know God is not a respecter of persons—what He did for them, He will do for us.

God wants to heal people. Everywhere He went, Jesus healed everyone who came to Him. When He sent out the Apostles the first time, this is what He did. "When He had called His twelve disciples to Him, He gave them power (authority) over unclean spirits, to cast them out, and to heal all kinds of sickness and all kind of disease" (Matthew 10:1). When He sent out the seventy He said to them, "Heal the sick, and say to them, 'The kingdom of God has come near to you'" (Luke 10:9). Before I left for the Philippines the first time, I drove to Pennsylvania to say goodbye to some friends and family. The first evening I was there I went to visit one of my closest friends and his family. That evening he, his wife, and there three young sons gathered around the kitchen table, and I was able to share the Gospel with them and they all prayed to receive Jesus. That Sunday, my friend and his wife took me to an open house that his employer was having. As we were leaving and driving down the long driveway from their house to the road, they mentioned to me that she, his wife, had cancer. They said that the husband was having some problems with their business, so she hadn't said anything to him about the cancer, because she didn't want to worry him. I told them that Jesus would heal her, so we stopped the car, turned around in our seats, stretched out our hands toward her and prayed. The Lord healed her.

On Saturday, after the Friday evening they had prayed to receive Jesus, I went and bought them a Bible. When I gave it to him, I told him to start by reading the book of John first, and then to read the New Testament. After I arrived back in Oklahoma, I called them and as he and I talked, I asked him if he was reading his Bible. He told me, "Yes, but that book of Job is hard to understand."

There was a Christian lady in Oklahoma that I was talking to one day. She was telling me that her middle-aged brother was terminal with cancer and was living with their mother. Their mother was elderly, and the Christian lady said to me, "My brother is such a baby. He won't do anything for himself. My mother has to do everything for him." So I said to her, "You know that God will heal your brother; let's pray for him." We prayed for his healing, and God healed him. James 5:16 says, "Pray for one another that you may be healed." God is waiting for us to tell Him what we want Him to do for us. When people are healed, they are blessed, and God is glorified. God is not pleased when we do not heal the sick. We read what He said to His ministers in Ezekiel 34:4; "The weak you have not strengthened, nor have you healed those who were sick, nor bound up the broken, nor brought back what was driven away, nor sought what was lost; but with force and cruelty you have ruled them."

When I first arrived in Baguio, I was asked by some of the employees at the Hyatt Hotel to start a Bible study there. During the first meeting, I gave an evangelistic message and prayed with them to receive Jesus. I attended a couple more meetings, and after a while, the employees took it over. About four years later, someone asked me, "Do you remember the Bible study that you started at the Hyatt?" When I said yes, he said, "It is still going on." Thank God for fruit that remains!

God's Protection

In 1990, there was a major earthquake in Baguio, and that hotel was completely destroyed, and many were killed. I'm thankful those employees had an opportunity to receive the Lord Jesus and be built up in Him before it happened.

I was in the United States during the earthquake. When I finally was able to talk to Naomi, the woman who took care of the ministry while I was gone, she told me about all the death and destruction there and the ministering they were doing. When she told me about the house I was renting (the one that I rented after Dada's and Tessie's), she said that it was heavily damaged, and all of my things were destroyed.

I thought, *Oh, Lord. How could my things be destroyed?* Before I left the Philippines for my first trip back to the United States, the Lord had given me Job 5:24:

> You shall know that your tent is in peace;
> You shall visit your dwelling and find nothing amiss.

There had been a severe earthquake after I left for the first time, and none of my things had been hurt except, according to Dada, a broken empty pop bottle or two under the sink. Quite frankly, I'm not sure they were broken during the earthquake.

A few days later, I was talking to Maribel. During our conversation, I made some mention about all of the things being destroyed. She said, "Oh, no. None of your things were damaged at all."

I found out later why Naomi had told me what she did. Because of all of the aftershocks, she was afraid to go into the house. And because of all of the other damages, she just assumed all my things had been destroyed.

As I said before, I believed the Lord was saying that it was time to go home. I had been there for three years. Shortly after this, I received a check from my home church for $3,000, so I began to make plans to return home to the states.

I spent the last night in the hotel. Before I left for the airport, I had lunch with the manager and his wife in one of the hotel dining rooms. When we walked into the lobby, there was a large group of my friends, waiting to say goodbye. As we parted, we stood in a circle and held hands and prayed. And as we did, we all wept.

When I got to the airport, there were two men from Baguio waiting to say goodbye. From Baguio to that airport is over a six-hour trip by bus. What a wonderful, wonderful ministry the Lord has given me! There is a song that says, "I left my heart in San Francisco." I left mine in the Philippines.

That trip back to the United States lasted one year and seven months. Even though I only ministered one time, the Lord fully provided all I needed to live comfortably for that period.

On that one ministry trip, I again saw the Lord's protection.

After I arrived back home, I contacted a friend, and we visited over the telephone. A short time later, he invited me to his church to minister. About that same time, my home church contacted me and said that they would like to provide a car for my transportation for a while. So I was set for the trip.

As I was reading my Bible the morning I was to leave, I came to Psalm 37. As I began to read, the first part of verse 1 seemed very alive to me, but I didn't know why—"Do not fret because of evildoers."

That night when I arrived at my destination, I found out why. That afternoon, as I was driving on a stretch of I-70, a carload of people were driving along the same stretch of road murdering people. The Lord spared me.

Then later, He spared me again. That May, I was making a trip to visit relatives. On the day I arrived, I called ahead to let them know what time I would be there. As I got close to their home, I missed a highway that I should have turned onto and went way out of my way. By the time I realized my mistake, it was dark. To save time, I found a state road on the map and decided to take it as a shortcut. As I was driving along (not speeding, I don't believe that the Lord can honor that) trying to make up time, all of a sudden the Lord spoke to me sharply, "Slow down." When He spoke, I hit the brakes. Then, I saw why. There was a sharp curve there, and at the same time, the road turned into gravel. I would never have made it. I would have gone off that mountain and been badly hurt or killed. If He had spoken a second later, it would have been too late. During my visit, the man that I was visiting said, "A lot of visitors who come here are killed because they drive too fast for these roads." I felt like saying, "Yes, I know."

When I returned to the Philippines, after that trip home, there was another time that I believe the Lord spared me.

There was an alcoholic man I knew in Dagupan City. He had been so abusive to his family that the chest of one of his sons was caved in a little, and I was told it was because of his father's beatings. The children would sneak out at night and steal fruit from their neighbors' trees because they had nothing to eat. It was so bad that the Philippine constables came to their home one day and asked his wife if she wanted them to take him out and kill him. She said no. Thank God she did, because he got born-again later.

After he got born-again, he continued to drink. He no longer abused his family, but he then abused himself. When he got drunk, he would do things, like break flowerpots over his head—flowerpots with dirt and flowers in them. I was so concerned about him and his family that I would pray for them often.

One afternoon, when he was drunk, he came to see me. He asked for money to go to another town where he had been working to pick up some clothes he had left there. I knew he would use the money for alcohol, so I told him no. He kept insisting and insisting, so I finally gave in and went into the house to get him some money. As I was walking up the stairs to my bedroom, the Lord asked me, "Who are you doing this for, you or him?" I knew I was doing it for me, because I just wanted to get rid of him. So I went back and told him no.

He had been there so long that he was starting to sober up. I told him that I wanted to help him if he wanted me to. I told him that I would try to get him a job, and he said okay. Mr. Santo Domingo, my landlord at Dagupan, was building a new apartment building, so I went and asked him if he would give this man a job, and he said yes. That was a great opportunity for him. Construction jobs are very valuable there and hard to find. The landlord also said that he could sleep in one of the apartments that was near completion. I gave him some clothes and told him he could eat his meals with me. He was also welcome to spend his spare time in my home if he wanted to. I also gave him a blanket to sleep in. The next morning when we went to get him for breakfast, he was gone. I later heard he had gone to Manila and went to work for a ministry there.

About four years later, his wife wrote from another town where she and her children had moved and asked if I could give them some financial help. Since I would be going so far, I decided to spend the day there passing out tracts. When I arrived, I went to the post office and found out the location of the house. I spent the rest of the morning passing out tracts, and at noon, I went to her house. When I got there, her husband was there, and he looked great. I didn't recognize him. He looked ten years younger. He was neat and clean, had put on weight, and had good color. He was obviously prospering in the Lord. I believe the Lord took me there that day to see him and encourage my prayer life.

After a short visit, I left and went back downtown and spent the rest of the day passing out tracts. That evening as it began to get dark, I was standing at the entrance of the market. I had noticed some men who were standing across the street at the entrance of the park, kind of looking at me, but I didn't think too much about it.

The market was busy, and I was able to pass out a lot of tracts. When the bus to Baguio stopped across the street to load passengers, I thought I should probably get on and go home, but there were so many people that I hadn't given a tract to yet. All of a sudden, it came to me very strongly, "Get on that bus." So I ran across the street and boarded the bus. I believe I must have been in some danger.

CHAPTER 7

Time to Return

In the spring of 1986, I had been home for well over a year and was waiting for a word from the Lord about my return to the Philippines. One Saturday morning, the Lord spoke to me and said, "Take the money you have and take [He named a local minister] and buy him some clothes to sow for your trip back to the Philippines."

When I called and told him I wanted to take him and buy him some clothes, I didn't tell him that the Lord had said to do it. I had about $500, so what I had in my heart was to buy him one good suit, some dress shirts and ties, and some other things he might want.

I picked him and his wife up and took them to a store that sells good men's clothes. When we got there, we found they were having a two-for-one sale on some of their suits, so we were able to get him two nice suits, the dress shirts, and neckties.

That afternoon before we left the store, he said, "By the way, when you're ready to return to the Philippines, we want to buy your ticket" (he meant their church). I told him that I would let him know when I was ready.

Later that day as I was driving home, the Lord spoke to me and said my rent would now be 4,000 pesos a month. Sure enough, before I left for the Philippines, I was notified that my rent had been raised to 4,000 pesos as of January 1, 1986. During 1985, I had paid the rent through various channels, but in 1986, it was decided for me to wait until I got back to settle up.

When it seemed the time was right, I contacted the man and told him that I was now ready for my ticket. He recommended that I contact a travel agency he knew about, so I contacted them; and they made my travel arrangements, scheduling me to leave for the Philippines on July 22, 1986.

In January 1986, it came to me that I would be returning on United Airlines. That seemed strange because I didn't know United Airlines flew into the Philippines. I thought I would check it out, so I called and asked if they flew to Manila, and the lady said no. As we were finishing our conversation, she asked if I would be flying after a certain date. When I told her yes, she said that they would be able to help me after all. She said that as of that date, they were taking over all of Pan American Airline's routes in the Far East, and Pan American Airlines was taking over all of their routes in Europe.

Sure enough, when the travel agency made my reservations, I was booked on United Airlines. Since I was not comfortable with flying, the Lord may have done that to strengthen me.

As previously, I continued to look to the Lord and made no arrangements or provisions for my needs. From various sources, I received enough money to buy the clothes and miscellaneous things that could not be bought in the Philippines. However, I received none of the finances I would need when I arrived.

First, I would give the tithe on the money I received. Then, I would have to pay Dada my back rent for 1986, the small electric bill that had accumulated, and 8,000 pesos for the food Chewie had eaten since I left. Before leaving, I had told Dada to keep track of his food cost, and I would pay it when I got back. She said no and that she would be glad to buy his food; however, I insisted, and I'm sure she was glad I did. No one expected me to be gone as long as I was.

A Lesson in Patience

The total I would need when I arrived in Baguio was about 50,000 pesos, which was about $2,500. Everything I needed was provided except that money. A friend even volunteered to take me and all of my luggage to the airport in his pickup truck. When I contacted Philip and Maribel, in Manila, to let them know I was returning, they volunteered to pick me up at the airport in their truck and take me to their home to stay.

When I got to the desk at the airline departure gate, I found that the flight had been oversold, and the airline people were trying to decide what to do. When they finished loading the plane, a few of us were still standing at the gate. If I couldn't get on that flight, I would lose all of my connections for the whole trip. I didn't say anything, I just stood there and listened and waited. Finally, they said I would be boarded on that flight. They would seat me in first class.

On the flight over, I was really concerned about where I was going to get the needed money. The thought kept coming to me, *I had better call Bob* [my brother] *and get the money from him*. However, I kept thanking God that He would provide everything I needed and put away all the other thoughts by meditating on what He has promised in His Word.

After I had been in Manila a few days and the 50,000 pesos were not forthcoming, I decided I had better call my brother and get the money. I felt I should go to Baguio and give Dada the money I owed her. I called Bob, and a couple of days later, I had the $2,500.

When I received the money, I went to Baguio and settled up with Dada. When Chewie saw me, he went bonkers. It was certainly a happy reunion!

It was the strangest thing. When I returned to the United States the first time and was sitting in the Dallas/Fort Worth Airport at seven o'clock in the morning, waiting for my flight to Tulsa, I saw a lot of business people catching flights out of town. As I sat there watching them, I prayed and thanked the Lord I wasn't doing that anymore. There is nothing wrong with that life if that's what the Lord has given you to do, but I was so happy and content with what God had called me to do and the life He had given me.

When I got back to the United States, it seemed like I had been gone a lifetime. It was no longer home. My family and friends still loved me, but the place I once occupied in their lives had been filled with other people and things.

But when I got back to the Philippines, it was exactly the opposite. It didn't seem like I had been gone over two weeks. I stepped right back into the lives of the family and friends God had given me there and right back into the ministry as if I hadn't been gone at all.

When I saw the jacket I had hung over the back of a chair before I left, it was hard to believe I had hung it there one year and seven months earlier. It was as though time had stood still. It was truly supernatural.

After I saw Chewie and settled up with Dada, I stayed a few more days in Baguio and then went back to Manila. Because of the heavy monsoon rains, the day I left was the last day I would have been able to get out of Baguio for a while. Right after I left, the last road to Manila was washed out.

On Sunday, some friends asked me to go to church and then to lunch afterward. After lunch, they invited me to their home. After we had visited for a while, I began to get weary and decided it was time to leave. As I was thinking about leaving, the Lord spoke to me and said, "Be patient." After we visited a little while longer, they handed me a check for 50,000 pesos, the money I had needed. If I had only waited!

Distributing Tracts

It was sometime after arriving back in the Philippines that the Lord directed me to start going to the schools to distribute tracts. Since the medium of instruction in Philippine schools is English, I put a tract written in the local dialect inside an English language tract. I would give one of these to each principal, teacher, and student. The English tract was for those at the schools and their family members who read English, and those in the dialect were for family members who did not read English.

When I first arrived in the Philippines, I had prayed for a van, but never received one. As the years went by, I saw why. Since I had no transportation of my own, I had to travel by tricycle (a motorcycle with a sidecar for passengers), jeepneys (originally jeeps converted into small buses), buses, and taxicabs. Because almost everyone in the Philippines travels by public conveyance, I was in constant contact with multitudes of people. As I traveled, I gave tracts to my fellow passengers. When I was in bus terminals, I distributed tracts to the people in the waiting rooms and to those sitting in the buses at the gates.

On every trip, I would distribute hundreds of tracts, and on some trips, I might distribute a thousand or more. There were periods when I traveled every day. Over the years, I was able to minister the Gospel to hundreds of thousands of people in this way.

One night when I was waiting for my bus to leave, I was going through the terminal passing out tracts. I saw a man sitting at a service window, so I went over to give a tract to him. As I did, he pointed down, and there I saw one of my tracts under the glass countertop.

The day I went to court for the hearing on my long-term (9-G) visa, I had to go into the judge's office first to finish some paperwork. I was in an outer office, and one of the clerks was going through the papers with me. When I looked down, I saw that he had one of the tracts under the glass top on his desk!

What an exciting ministry! And to think, I was unhappy when the Lord gave it to me! For a few years after He gave me the tract ministry, I was disappointed. I thought I was going to hold open-air crusades. I saw the tracts as a second-class ministry.

My Tract Ministry

I wanted to get out there and do something. I was not excited about this ministry at all. Then one day, the Lord said to me, "Tom, every Word of God that is ministered in this world comes off the printed page." I thought, *Yes, that's right*. All of a sudden, I saw the tract ministry in a whole new light.

Later, a woman read me two accounts of outstanding men of God who were born again through tracts. When she did, I wept. My heavenly Father knows what I have need of.

Because I was going to a lot of schools in the countryside and didn't have my own transportation, I had to walk. As I walked these roads, I would give a tract to everyone I saw, and I would leave one at every house I passed unless it was so far off the road that it was impractical.

When I got to a school, I would stop at the office and ask permission to go into each classroom and distribute tracts. Out of all the public schools I went to outside of Manila, only three would not give me permission.

At one of those schools, I distributed tracts to the students as they left for lunch. At the second, I went early in the morning and gave tracts to the students as they arrived. The third was a high school with about 8,000 students. It had a high wall around it with one main gate for entry, so I invited a number of people from a church in the area to help me give tracts to all the students as they arrived.

I went to public elementary schools, high schools, colleges, universities, and religious schools. Also, as I traveled to the various towns to go to their schools, I would spend one day or several days, depending on how large the town was, distributing tracts on the streets and in their market. Each town has one or two days a week they call market day. This is the day when the people from the outlying areas come to the town market to buy and sell. Market day is the day I would try to go to towns. That way, not only was I able to reach the people in that town, but also I was able to reach the people from the outlying areas. How I love traveling around the countryside passing out tracts and being in contact with all those precious people!

The Tract Ministry Expands

Many other people were getting tracts from us and distributing them in other parts of the Philippines. Others were taking or sending them to other parts of Asia and islands in the Pacific Ocean. One way the Lord set this up was through a woman in Manila who has a ministry of intercession. When ministries came to the Philippines to hold meetings, she would go and offer her services, and because of that, she became well-known. After a while, the Lord told me to set up a tract distribution center at her house. Through this means, many ministers who came to Manila for meetings and other reasons were able to pick up tracts and take them back to their home areas for distribution.

There was a time when about twenty policemen were being brutally murdered in Manila, a month. While the policemen were standing or walking on the street, men would come up behind them, shoot them, and take their guns. When I heard about that, I was grieved in my heart—first in concern for the policemen and their families and then in outrage toward those who would shoot a policeman in the back for his gun. I began to pray for the policemen in Manila.

One day, a Christian policeman came to a place in Manila that was distributing our tracts. He said that he wanted tracts to give to the police officers in Manila. Praise the Lord! He gave us an opportunity to minister to those policemen. Shortly after that, I didn't hear about any more policemen being shot.

How Precious Are You?

You may wonder why, from time to time, I refer to people as "precious." After I had been ministering in the Philippines for a while, the Lord asked me, "Tom, how precious are you?" I didn't know how to respond to that, so He said, "How serious would it have been if you hadn't been saved?" That I could relate to!

If I had not been saved, it would have been a catastrophe. Then, He said, "As precious as you see yourself, see everyone else." So, with God's help, I am now able to see everyone as precious, and they are very precious, even the men who shot the policemen.

Another person who got tracts from us also blessed me. The hotel where I stayed is on the shore of Manila Bay. Most of the rooms I have stayed in face the bay. When I looked out the windows, I could always see big ships anchored there, loading or unloading cargo.

Then, the Lord put all the seamen who were on those ships on my heart. I began to realize how badly they needed Jesus, not only so they could receive salvation, but also so He could live in their hearts and help them day by day.

I knew that these men had little opportunity to receive ministry or Christian fellowship. And secondly, they are separated from their families for long periods of time, and I'm sure they get very lonely.

Evangelizing Seamen

Because of my great desire for them to receive Jesus, I began to think about how I could minister to them. I thought about going on the ships, but I realized that was impractical. I also realized, because Manila is a big city, once they got on shore, there would be no way to find them.

Since I wanted to reach the seamen, when I was in the large market area and the streets near the port, I always kept a few English tracts on the bottom of the national language Tagalog tracts in my hand. That way, when a seaman came by, I could quickly give him an English tract off the bottom; however, I was not able to reach many of them that way.

Then one day, a ministry in Manila that distributes our tracts contacted us with a request for a large number. They said that a man came by and requested 110,000 tracts, because he had the names and addresses of all the Filipino seaman in the world, and he wanted to send each of them a tract. Again, praise the Lord. I believe that those tracts reached not only the Filipino seamen but also a lot of other seamen throughout the world.

If each of us will just do what the Lord has called us to do, we will reach the multitudes of this world!

A Change of Heart

It is interesting how the Lord helped me to start distributing tracts. After He gave me the tract, I began to print them and give them to whomever wanted them for distribution, but I was uncomfortable with the thought of handing them out myself. While I was living in Dagupan, a couple came to visit. While they were there, they invited me to drive to Baguio with them. I had never done it before, but that day, I decided to take tracts along.

When we got there, they decided to go to the market and shop. I really didn't feel like going, so I told them I would stay in the car and wait. As I sat there, I started to feel uncomfortable. Here I was with all these tracts, and there were thousands of people who needed Jesus, streaming past the car, and I didn't have the boldness to get out and give them a tract. I was ashamed of myself.

All of a sudden, I heard someone say, "Brother Tom." When I looked, I saw a group of young people from a church that I knew. I got out of the car, and we began to talk. As we talked, I began to think about the tracts. I asked, "Would you like to pass out some tracts?"

They all excitedly said, "Yes!" So I got the tracts out, and they began to give them to the people passing by.

After a while, I took a few tracts and gave them to some people passing by. I realized, *This is not so bad*. I took some more, and as I continued to hand them out, I realized I actually enjoyed it. From that day on, I have enjoyed distributing tracts as much as anything else I have ever done. It is fun! In the Philippines, from what I have observed, fewer than one tract in a hundred is thrown away. People keep and read them.

When I was in Manila on weekdays, I went to the schools, arriving early to give tracts to the teachers and students as they entered. On weekdays and weekends, I would also go into the streets and shopping areas and distribute tracts.

Something Good from America

One morning after I left an elementary school, I began walking down the street to a busy thoroughfare giving tracts to those I met. Just as I got to the thoroughfare, I saw a large building. When I stepped inside, I saw a large room full of young women and a few men. Some of the women were scantily dressed, so I thought I knew what kind of place it was. I gave tracts to all the girls and some of the men and then left. When I got to the main street and saw the front of the building, I knew my suspicions were right.

As I stood on the thoroughfare passing out tracts, I looked up and saw large billboards that were advertising terrible, wicked movies; and I could see that one of them was American. The billboards had large, graphic scenes of what the movies depicted. As I stood there looking at those billboards, I said to the Lord, "Lord, doesn't the United States have anything better than that to send over here?" The Lord answered, "You're here." How encouraging it was to hear that!

The day I started to distribute tracts in the schools, I selected an area in the lowlands at the foot of the mountains.

In the countryside, concrete markers are set up where country roads that have schools on them intersect with highways. On these markers are the names of each school on that road and the distance in kilometers each school is from that intersection.

The marker at the road that I picked listed four schools. The distance to the farthest school was a little over seven kilometers (about four miles).

Since I was new at this, I had a lot to learn. The first thing I learned was to utilize all of my time each day. The daily goal I set for myself was four schools, two each morning and two each afternoon. Sometimes, I would only reach three, but if they were on a highway where I didn't have to walk, I could reach five or six.

As I said, one of the things I had to learn was to utilize my time. On the first morning, I didn't leave home until after seven o'clock. That meant I didn't arrive at the road to the schools until about nine o'clock.

When I went to schools outside of Baguio City, I would take a taxi to the bus station and then take one or two buses to the area where I was to work that day.

It was vital to start my day by spending quality time with the Lord, reading and studying the Bible and ministering to Him. Without consistently starting my days with this devotion time, I don't think I could have made it. It was during this time He strengthened and enabled me to do what I did.

I would get up at three o'clock, shower, have my devotion time, and leave the house at about 5:45, which allowed me enough time to get to the bus station and catch the six o'clock bus or sometimes a jeepney, depending on where I was going. I spent most of my travel time quietly ministering to the Lord in praise and worship and in praying in my understanding and in the spirit for the ministry.

Actually, it was about 9:15 when I got off the bus that first morning. I saw on the marker the first school was one kilometer away, the second three, and the third five. I was told the Philippine government does not want the children to have to walk over one kilometer to school if possible. I think it is great they are that concerned about their children.

It was already late, but I thought I would still try to reach the first two that morning. Just going to the schools is not where most of the time is consumed; it's going to the houses along the roads as you go to the schools.

Demonic Oppression

When I went into the first school, it was full of demonic oppression. The people and the place seemed devilish. The devil often does this sort of thing when we are beginning a new area of ministry. I guess he was trying to intimidate me from starting this new work, because it never happened at any other school.

After I left the first school, I realized that if I continued to go to the houses, I wouldn't reach all the schools, so I quit going to them. When I got to the second school, they had dismissed for lunch minutes before. That meant I had to wait about two hours until the afternoon classes began, so I laid down in the shade and took a nap. When I woke up, I was surrounded by little faces, staring down at me.

After I left the second school, I rushed so I could reach the last two schools that afternoon. I walked as fast as I could. It got later and later, and it was also very hot. I began to get frustrated, and then, my frustration turned into discouragement as I began to realize the best I could hope for was to reach one more school. Then all that frustration and discouragement turned into tears, and I cried out to the Lord that I couldn't do it; I wasn't going to make it. I felt like quitting, but I kept putting one foot in front of the other.

I finally reached the next school, which was in a small village. By that time, it was close to three o'clock. The schools in the Philippines dismiss at 3:30, so I would have enough time to distribute tracts in this school, but that would be all.

When I finished that school, I asked about the next one. They told me this was the last school, that this village was the end of the road. I don't know how I missed the third school as I was looking diligently for it. I don't know if it was so far off the road and unmarked that I didn't see it or maybe it had been closed down. It may have been closed down because the Lord did not send me back there.

After I distributed tracts in the village, I began walking down the road back to the highway. Soon, some men in a jeepney stopped and asked if they could give me a ride. I jumped in and off we went. This was great and the air blowing on me was very refreshing, but as we began to pass all those houses I had not stopped at, I realized I could not ride past all those people and not give them the Gospel. I asked the men to stop and let me out, and I went door to door back to the place where I had quit going door to door that morning. Just as I got back to the highway that evening, the Lord spoke to me and said, "Well done, good and faithful servant."

He, of course, gets all of the praise, honor, and glory, because it is by His grace that we do all that we do.

> But by the grace of God I am what I am, and His grace toward me was not in vain; but I labored more abundantly than they all, yet not I, but the grace of God which was with me. (1 Corinthians 15:10)

But it was certainly good to hear those words of appreciation!

Weep with Those Who Weep

Once when I was going through a situation that broke my heart, I began to weep. As I anguished about it, I sensed the Lord saying to me, "As I have told you to weep with those who weep, I do the same with you."

The same was true as I walked those hot, dusty roads, trails, paths, rice fields, and forests. It was so hot and humid that by ten o'clock on some mornings, I would have to take the paper money out of my pockets and put in my bag or it would get soaking wet from perspiration. In the mornings when I left Baguio, it was cold, sometimes close to forty degrees; but I couldn't wear more than a light shirt and a light jacket, because of the heat I would experience later in the day. The reason the jacket had to be light was so I could roll it up and put it in my bag, so for the first hour or two, it was freezing.

Although there were times I experienced physical discomfort, I really enjoyed the ministry to the schools. This is because of the Lord's presence. It was wonderful to be able to spend my days walking with Him. When someone else was with me, I found myself resenting their presence, because it hindered me from sensing His presence.

Then, there were the times He would say or do something. They were added bonuses. No matter what my flesh was going through, being with Him and working with Him was so great that I hardly noticed how my flesh felt.

An Opportunity across the River

The ministry the Lord gave me is somewhat different from what you would normally think a missionary does. In one town, I was told about a group of people who were there with a tent and who taught the same thing I did. When I went to visit them, I found they were a large group of Filipino, Italian, and American evangelists who traveled throughout the Philippines. It was a great work. The man in charge said that he had been traveling like that for nine years, and in all that time, he had never seen anyone else out like I was.

Once when I was in an elementary school classroom, the teacher said to me, "The reason the children are looking at you like that is because they have never seen a white person before." So when I'm out evangelizing, I'm an unusual sight.

As I was finishing one area, I was told there were schools on the other side of some low mountains. I would, however, have to cross a river that had no bridge to get to the mountains.

On the morning I went there, I asked some of the local people how I could cross the river, and they told me to follow a certain trail. When I got to the river, I found some men with bamboo rafts transporting people back and forth across the river. They used long poles to push the rafts along.

After I got across the river, some of the other people who had also crossed and I began walking toward the mountains. As we walked along, I heard a jeepney coming from behind us, driving along the wide, sandy riverbank. It came up to us and stopped, and those people began to get in. It was already jam-packed, but I got on too. I stood on the back step and held on, and some of the people inside reached out and held on to me. After leaving the riverbank, the jeepney went through a small community and then drove up into the mountains. After we had gone as far as the jeepney could go, the driver stopped, everyone got out, and we started to walk again. As always, I was asking questions as I went to find my way.

By the time I got to the first school, a high school, they had already dismissed for lunch. As I questioned the people there, they gave me the location of their elementary school. Since it would be about an hour and a half until afternoon classes started, I thought I would walk to the other school. After I walked a ways, I came to a stream. After I followed it for a while, I found a shallow place, took off my shoes and socks, and waded across.

When I got to the school, they were still having their lunch break, but all the teachers were there. Again, I didn't want to lose time, so I asked the teachers if they would distribute the tracts to the students if I gave them to them, and they promised they would. After I counted out the tracts and gave them to the teachers, I walked back to the other school. By the time I got there, classes were in session. I went into the classrooms and passed out the tracts and then started to walk back out of the mountains.

As I walked, I was told about two more schools. One I did not think I could reach that day, but the other one was to the east at the bottom of the mountains, so I thought I might be able to get to it before they dismissed. I followed the directions I received, but could not find it anywhere. When it got to be the time they would be dismissing, I decided to come back another day and started walking to the river.

I was so tired, I lay down to rest, and fell asleep. I awoke rested and continued on. By the time I got to the river, it was starting to get dark. About a 30-minute walk on the other side of the river was a highway that led up into the mountains to Baguio. Once I caught a bus at the highway, it would be about an hour-long ride to Baguio, home, dinner, a hot shower, and bed.

As I got to the river, I saw a young man standing on a bamboo raft out on the water. I called to him and said, "Will you take me across?" His response was, "Go around, Born Again." Then, he poled away. A river with no bridges in the area makes it a little hard to go around. This was another opportunity for me to walk in love. I've learned that no matter what, walk in love, be quick to forgive, and be patient. If the provision we think we have is not working out, it's not from God. He has the ability to make His provision manifest.

One thing about being a red-headed, Caucasian evangelist in that part of the world is that everyone knows you're in the area and what you're doing. The Gospel is preached in a lot of different ways.

Even though the man on the raft wasn't happy about me being there, he had heard the Gospel.

I sat down on the riverbank and wondered what I was going to do. The place was deserted; there were no other men with rafts nor anyone else in the area. It was almost dark, but down the river to my left, I saw some lights. I thought it was probably the small community we had passed through that morning. Since Filipinos are such hospitable people, I thought if I went to one of those houses and asked, they would probably let me spend the night and give me something to eat. I did not want to spend the night out if I didn't have to. I didn't have a mosquito net, toothbrush, etc. I didn't, and God doesn't, want us to suffer unnecessarily.

As I sat there looking across the river, I saw an elderly man walking on the other side, through the underbrush, toward the river. I saw him, but I didn't pay much attention to him. He had a bundle of sticks that he was carrying on the top of his head, probably taking them home to cook his dinner.

As he kept walking toward the river, I began to watch him more closely. He walked up to the river, then down the bank, and into the water. He had on a pair of pants that were cut off about four inches above his knees, and as he continued to walk down into the river, I could see the water rising higher and higher on his legs. I saw his knees disappearing as he walked deeper and deeper into the water. By then, I was watching him intently. As I watched him, I noticed the water level began to move down his legs, and then, his knees began to appear. The water was getting shallower, and as I watched, he walked up out of the river and disappeared into the bushes.

Did I ever rejoice! This was a ford. When the man poled away, I did not know how long his pole was, so I couldn't tell how deep the water was. I just supposed it was the same depth as where I had crossed that morning.

I believe with all my heart that the Lord sent that elderly man at that time to show me how deep the water was. Besides that, had I walked down to that river a hundred yards to the east or west, I would not have seen him. The Lord had led me straight to that ford. I took my shoes and socks off, rolled up my pant legs, and waded across the river. Actually, the ford was a place where jeepneys cross the river. When I got on the other side, I found a road and followed it to the highway.

The day I came back to go to the other two schools, I got off the bus at the end of that road and started walking along it toward the river. As I walked, a jeepney came along and picked me up. After he crossed the river at the ford, he turned west and drove along the dry sandy bank of the river for some distance. You couldn't see tire tracts in that soft dry sand. That was why I couldn't tell it was a ford the previous night.

I spent the rest of that day going to the other two schools (one was closed) and the houses in the area. When I got back to the river that afternoon, I began to think of all those sharp stones in the riverbed that I would have to walk over. When I had crossed the previous night, those stones had hurt. Also, when I crossed a stream or a river in bare feet, I would have to walk up a muddy or sandy bank with wet feet and then put my socks and shoes on with my feet in that condition, which was uncomfortable. I didn't want to go through that again if I didn't have to.

When I got to the river, I sat down under a tree and prayed. I asked the Lord if a jeepney would be coming by, and He said, "Yes." As I sat there waiting for the jeepney, it started to rain, so I was glad I was under the tree. As the time passed, I started to wonder if what I had heard was God, so I prayed again. I said, "Lord, did I hear you, is a jeepney coming by?" Again, I got a yes, so I continued to wait. After a while, a jeepney did come, but it was going the wrong direction.

As I continued to wait, some women walked down the riverbank and started to cross the river. Just as they started to cross, I heard the sound of a motor from down the river on my left. When I looked, I saw a jeepney coming toward us. When he got to where I was, I flagged him down. and he picked me up. As we passed those women wading across the river, I had a strong urge to say to them, "You should have waited for the jeepney," as if I knew something they didn't. I guess that is spiritual pride.

An Open Door in an Ancient City

On another day, the Lord told me to go to a certain religious high school and pass out tracts. I knew He meant at the gate, because they would never allow me to come into their school and distribute tracts unless He somehow opened the door. I had been in that area that day going to other schools when He spoke to me.

When I checked in my bag, I had about 1,000 tracts left, and I hoped that would be enough. It's amazing how the Lord does all these miraculous things, yet I hoped I would have enough tracts.

When the Lord tells us to do something, we can rest assured all we need has been or will be provided. He could have multiplied the tracts, if He had wanted to! Not getting a job done is never because of a failure on God's part.

The town I was in is named Agoo (pronounced Ago-o). A marker along the highway as you enter town gives the date when the town was established. I don't remember the exact date, but I do remember that it was before 1492, the year Columbus discovered America!

I knew the school layout well, because I had passed it a number of times. The school building sat back off the street about 125 feet. The building itself was about four stories high and was built like a "C" with the open side facing the street. The inside of the "C" and all of the way to the street was concrete. Across the street was their religious headquarters. I wondered how I could stand there and pass out tracts without being seen by religious or school officials. If they saw me, I knew they would make me leave.

There was only one gate. When school was dismissed, those students would stream out that gate a lot faster than I could give out tracts. I knew from experience that about 1,000 tracts an hour was the fastest I could hand them out. I also knew those students would be out the gate and gone in ten to fifteen minutes. As I thought about this, the Lord spoke to me and said, "The students will leave slowly enough for you to give them the tracts."

I stood in front of the gate and waited. After a while, some men came out and put up a volleyball net across the section inside the

"C." Shortly after that, the students were dismissed, but they didn't leave the grounds. They began lining the porches and concrete area. As I continued to watch, two teams of girls came out and began to play volleyball. After that, two teams of boys came out and began to play basketball. It seemed to be a tournament.

Just as the Lord Had Said

After they had played a while, the students began to leave slowly, just as the Lord had said. Because of all the excitement, no one noticed me or what I was doing except one teacher. She came to the fence and asked what I was giving out, so I gave her a tract. As soon as she saw it, she said to some students nearby, "Don't take that." When she said that, they really wanted a tract!

The students left so slowly that afternoon that I had an opportunity to give a tract to each of them. Only twenty or thirty wouldn't take one, but the rest did. After the teams quit playing and most of the students had left, I walked inside and gave tracts to the few students who were standing around talking except for three or four boys who were playing basketball. As I stepped out on the street, I had one tract left. I saw an elderly man walking toward me, so I gave the last tract to him.

Everything had gone smoothly and efficiently that afternoon. Everything I had thought would be a problem, the Lord had already worked out. He had either set up the games for the day I would be in town or He had me in that town on that day. Years before, He had quickened Proverbs 3:5–6 to me, and I again saw that scripture fulfilled:

> Trust in the Lord with all your heart, and
> lean not on your own understanding;
> In all your ways acknowledge Him, and He
> shall direct your paths.

CHAPTER 8

Witnessing in Public Schools

The way the Lord got me into the public schools in that area is interesting. It is a strongly religious area, and I expected problems getting into those schools.

On the morning I went to the large downtown public elementary school, they were having a special program for principals and other school officials in that area. School was in session, so I went to the office and was given permission to go into the classrooms. After I had finished, I went to find the principal to thank him.

When I found him, the program had just ended, and they were all sitting down for refreshments. As we talked, he invited me to join them. He just didn't seat me at a table; he sat me with him as if I was a special guest. Then, he introduced me and treated me like a visiting dignitary. I believe it was because of him that I had no problems getting into the schools in that district.

One time when Daniel and his wife, Jean, came to visit, he said that they had been unable to get into the elementary school in their town, so I said I would go there one day and see if I could help them.

I had to leave early in the morning on the day we had chosen. It was the spiritually darkest and the most oppressive morning I have ever gone out to minister. The atmosphere seemed to be permeated with fear.

During the trip, an older woman got in the jeepney and sat beside me. She appeared to be a housewife and was very pleasant. As we rode along, we visited about various things and had light conversation. When we arrived in Daniel's town, the jeepney drove down the street the elementary school was on; so when it passed the school, I asked the driver to stop, and I got out. The woman also got out.

Unusual Permission

Daniel was waiting for me, and we walked into the school together. The principal was not available, so we had to talk to another man. When we told him what we wanted, he couldn't or wouldn't give us permission. He said that we would have to wait. Then, I noticed that the woman I had met was in the office and was talking to various people. After a few minutes, she told the man who had talked to us to give us permission, and we were cleared to proceed. I don't know who she was, but she seemed to have a lot of authority. The Lord had again given us the favor we needed!

> For You, O Lord, will bless the righteous;
> With favor You will surround him as *with* a shield. (Psalm 5:12)

As you can see by now, being in the ministry and being an American is not what you need to get the job done.

> Unless the Lord builds the house, They labor in vain who build it; Unless the Lord guards the city, The watchman stays awake in vain. (Psalm 127:1)

One afternoon, Digna took the students to a nearby town to pass out tracts. When they arrived, they found the town was having a fiesta that week, so it was full of people. While they were passing out tracts, a religious procession moved through town to a large cathedral, and the students followed. This church is part of a religious group that believes in Jesus but had been hostile to the born-again message. They had even placed articles in the newspaper against this message and those who preached it.

As the cathedral was filling with people, the students moved through the crowd outside. While they were passing out tracts, a priest came out and wanted to see what they were giving to the people. When they gave him a tract, he read it, and then, he asked them to come inside and give tracts to all the people. The Lord can open any door He chooses.

For we walk by faith, and not by sight.
(2 Corinthians 5:7)

Buying Bibles

Then, the Lord gave us a Bible distribution ministry. Since the time I fully committed my life to the Lord, I have had the desire to give Bibles to anyone I saw who needed one. I also enjoy donating money to ministries that distribute Bibles.

In the Philippines, there are few Christian bookstores. Once you leave Baguio, you have to travel at least thirty miles before you can buy a Bible. If you travel north from Baguio, you have to go a lot farther than that.

One time, the Lord put it into my heart to buy some New Testaments for the congregation of one of our first graduates, Juhn. While I was at the Philippine Bible Society buying them, I saw a paperback book that taught who Jesus is. It looked good, so I bought it and slipped it into the side pocket of the shoulder bag I carry my tracts in.

Juhn's church is about a hundred miles from Baguio, so a round trip takes almost all day by bus. When I asked directions in his town to his church, the people there referred to him as "Born Again"—the same name as that man who was on the raft that night had referred to me. They do not mean it as a compliment, but as we continue to preach the Gospel and lives are changed, that will change. That was 1988. In 2004, when I was visiting Juhn's church, he told me that there were more than sixty churches in that town, at that time. As we obey and serve God, He is able to do wonderful things.

When I arrived in town, the oppression was heavy. I found a jeepney that was going past Juhn's church, so I loaded the Bibles in the back and got in the front seat. Then, a woman from a cult got in beside me, and for most of the trip, she kept up a continuous harangue about the doctrines of her group. What a relief when she finally got out!

Dealing with a Cult Leader

When I arrived at their home, I gave Juhn and Ligaya the Bibles, and after visiting for a while, walked back to the main road to get a jeepney back to town. There was a young Christian man there visiting, and when I left, he said that he would go back to town with me.

When we got to the main road, I went to the houses there and distributed tracts to the people in them. I also gave tracts to a group of men standing at a waiting shed.

All of a sudden, a young man approached me and asked with great hostility, "Who told you, you could do this?" From where I was standing, I could see a building that a certain cult in the Philippines calls a "church." From the way he was dressed, talked, and acted, I supposed he was from that group and probably was the overseer in that area.

Then something happened I had not experienced before or since. I just looked at him. I did not respond verbally; I just looked at him. As he continued to talk, I continued to look at him. Then, he began to get nervous and shaky and began to say, "What are you looking at? What are you looking at?" But I just kept looking at him. Finally, the young man who was with me said something that broke this man's contact with me, and he withdrew. I know he was glad to be out of there.

It reminded me of Paul and the sorcerer in Acts 13:6–12. It says that Paul looked intently at this man and then spoke to him. I did not speak to this man; I just looked intently at him. I believe it was the Lord in me withstanding the spirit in him. It was an extremely unusual situation.

I don't know everything that happened that day, but I do know that young man found out that we Christians have a power in us they don't have. My prayer is that this experience will help to lead him and others to salvation.

Later, I was told that some friends were praying at the exact time this was going on. When we get to heaven, we will find out that many of our great victories were accomplished either through or with the assistance of the prayers of others.

> Who delivered us from so great a death, and does deliver us; in whom we trust that He will still deliver us, you also helping together in prayer for us. (2 Corinthians 1:10,11)

> For I know that this will turn out for my deliverance through your prayer and the supply of the Spirit of Jesus Christ, according to my earnest expectation and hope that in nothing I shall be ashamed, but with all boldness, as always, so now also Christ may be magnified in my body, whether by life or by death. (Philippians 1:19–20)

Bibles for Schools

When I bought the Bibles for that church, a woman I knew allowed me to use her discount at the Philippine Bible Society. Shortly thereafter, a Christian women's group she belonged to donated 400 New Testaments to me. The school children were really on my heart, so I decided I would distribute the Bibles in schools.

I selected an elementary school and set a day to visit. Since it was the first day of this new ministry, I invited some other ministers to go along to celebrate this new opportunity the Lord was giving us.

During my devotion time the morning we were to go, the enemy tried hard to intimidate me into not going. He is afraid of the Word of God and of our getting it to others. Fearful thoughts began to try to enter. When the enemy could not intimidate me personally, he tried another tactic with strong, oppressive thoughts about harm to other members of the group, that we would be taken captive by the NPA, and of harm coming to the women.

For a while, I really had a battle, but the fight of faith is a good fight, and through the Lord and the armor of God, I took those thoughts captive.

> For the weapons of our warfare are not carnal but mighty in God for pulling down strongholds, casting down arguments and every high thing that exalts itself against the knowledge of God, bringing every thought into captivity to the obedience of Christ. (2 Corinthians 10:4–5)

Before we left, I prayed and asked the Lord how many Bibles we should take. Not only did I want to be sure to take enough, but also I didn't want to take too many and then have to bring them back. The Lord said, "220." On the day I had been at that school passing out tracts, I had met an elderly man who had asked for a Bible, so we were also taking along a complete Bible for him.

This man had told me he was born-again, but I had not thought to ask if he had been baptized. Because of that, I took some towels and extra clothes along in case he would like to be baptized while we were there. He could not speak English very well, so we were taking him a Bible in his own language. I also thought that since Digna would be with us, she could help teach him about baptism in his language.

As we waited for our bus, a tricycle driver who was waiting for passengers came over and asked me what was in the packages. When I told him Bibles, he became excited. He told me that he had always wanted a Bible and asked if I would please give him one. I told him no; we needed every one of them, but he continued to ask and ask, so I finally gave him one.

When the bus got there, we loaded up and rode to the next town. There we took tricycles for our ride to the school, which was in the foothills west of town. When we arrived, we explained to the head teacher what we wanted to do, and she gave us permission.

One Bible Short

As we distributed the Bibles in the last classroom, we came up one Bible short. A little boy just sat there looking at us and then looking at the other children who were happily looking through their Bibles. I was quite unhappy with myself, because I had given his Bible to the tricycle driver!

The teacher said that it was all right because he had a sister in that class, and they could share hers. But that was not good enough. As I stood looking at him, I remembered the book about Jesus I had bought at the Philippine Bible Society weeks before and had forgotten about. I went to my bag, got it out, and gave it to him. I told him the book was about Jesus, and it was a special book just for him. It was a special book the Lord had provided for him. I told him that he could also read his sister's Bible.

The Lord had foreseen this situation and had provided that book for him. It was amazing! Any of those classrooms could have been the last one, and any student in that classroom could have been the last one, because we were not working by any kind of system. We were going into the classrooms as we came to them and giving the Bibles to the students as we came to them. In fact, this boy was not the last student on the last row. When we came up short on his Bible, the students all around him already had theirs.

The student we didn't have a Bible for was the one who had a sister in the same class. How many children have a brother or sister in the same class?

Also, I had prayed long before school opened, and the Lord already knew how many students would be in school that day. Actually, He knew a lot longer than that how many students would be there. Isn't He wonderful?

> Remember the former things of old, For I am God, and there is no other; I am God, and there is none like Me, declaring the end from the beginning, And from ancient times things that are not yet done, saying, "My counsel shall stand, And I will do all My pleasure." (Isaiah 46:9–10)

When we went to the elderly man's house, he told us that he had already been baptized, so we gave him his Bible, had a nice visit, and left.

A month or so before that, a woman from the United States had mentioned that she wanted to donate some Bibles to a Bible school in Baguio. Since I knew it would be hard for her to get there, I volunteered and asked what version they used. She said that she didn't know, so I told her I would check with the school and find out before I bought the Bibles, but I was busy and kept putting off going there.

Shortly after we went to that elementary school, I decided to take the rest of the Bibles to a high school. When I went there that morning, I didn't have enough Bibles for everyone in the last room, so I asked the teacher if she would make a list of the ones who did not get a Bible, and I would bring them one on another day.

Mister, Where Is My Bible?

One night, a few months later, as I was coming back from Manila, the bus stopped at a bus stop near that school. In the Philippines, there are vendors at many of the bus stops. They get on the buses and sell snacks—fruits, rice cakes, ears of boiled corn, etc.

When the vendors got on that night, there were two girls, probably about thirteen or fourteen years old, among them. I didn't pay much attention to them, but when they got to me, one looked at me very sternly and said, "Mister, where is my Bible?" She was one of the students who had not received her Bible that day. She had remembered, but I had not. It seemed to me, and I'm sure to the Lord also, that if anyone forgot, it shouldn't have been me.

Believe me, no Filipino girl would ever speak to a foreign man, and if she did, it would not have been in that tone of voice. I knew it was the Lord speaking to me though her!

That day, after I had finished passing out the Bibles, I walked across the road and waited for a bus back to Baguio. One usually comes by every five to ten minutes. As I stood there, a bus came by. I waved, but it kept on going. That was strange because they always stop. I continued to wait, and another bus came along. I waved at it, but the driver never even slowed down. This was very unusual! Another bus came by and then another. Finally, the fifth bus stopped, and I got in by the door that was about halfway back on the side of the bus.

When I get on a bus, I immediately give everyone a tract. Then, as new people get on, I give them one also.

When I got on this bus, the first people I gave tracts to were a young couple sitting in front of the door where I had entered. As I went on by them, the man called and said that he wanted to talk to me. I told him to wait until I had given tracts to the other passengers and I would be back. When I got back to them and we began to talk, he told me that he was the pastor at the Bible school in Baguio that I was supposed to go to. At the time, I thought this was great. It was saving me a trip to the school and those students were going to get their Bibles a lot sooner than otherwise. However, I realized later that it was the Lord's way of telling me to slow down. I was moving too fast. I wasn't taking care of business, as I should.

I told him about the Bibles the lady wanted to donate and that I needed to know what version they used. He told me, and soon after that, I took their Bibles to them.

A Change in Plans

Something else interesting happened one day. I once left a school on that road and was waiting for a bus. As I stood there, I saw a delivery van parked at a small store nearby. The people in the van were going to the stores along the road making deliveries.

That stretch of road up to Baguio is quite desolate. There is rarely a house, and the only other buildings are a few small stores that serve people traveling to Baguio and the people living in the surrounding mountains.

When the people in the delivery van saw me, they asked if I would like a ride. This is very unusual; it never happened before or after. First, these types of people are quite shy of foreigners. They also know how often the buses come along and how much faster I would be in Baguio riding by bus rather than riding with them.

The other thing that was unusual was that I took them up on their offer. Riding with them would take me hours longer to get back to Baguio, and I knew it. However, I got in and off we went. As they stopped at these little stores, I gave tracts to everyone.

I finally realized it was the Lord. I would never have walked that road and given tracts to those people, and the buses do not stop at those stores, so I could not have reached them that way. The Lord wanted me to minister to the people along this road, and He had provided a way for me to do it!

Now, I was out of Bibles. Once a man in the ministry told me that through Christ for Greater Manila (CGM), I could get New Testaments for about one-third of the price I was currently paying. Not only that, but also they provided excellent Christian teaching materials and course completion certificates free with the Bibles. After talking to him, I thought about going to see them, but I was busy, so I put it off.

A short time later, as I was passing out tracts in Manila, I gave one to an American, and he stopped, and we talked for a while. As we talked, he told me that he was associated with CGM. When I mentioned to him that I wanted to go by there someday and talk to them about Bibles, he told me that I would have to wait and that the man I would have to see was in the United States and would not return for a few months. Then, he gave me the man's name. I had been saved a trip.

About the time this man was supposed to be back, I thought about going to see him, but I again put it off.

A Divine Appointment

One day, while on a bus to Manila, I fell asleep. When I woke up, we were stopped at a bus stop. As I looked out the window, I saw another bus parked there. I thought I had better hurry up and give them their tracts before they left.

I got on the bus, and as I walked down the aisle distributing tracts, I came to a foreign woman with two small children. The children were three to four years old. When I walked back past her, she spoke and asked what organization I was with, and we began to talk. As we talked, she told me that her husband was with CGM. When she said that, I told her that one day, I wanted to go by their office to see the man in charge of the Bible ministry, and then, I mentioned his name.

When I said that, she said, "He's in there [the restaurant] right now having lunch with my husband. You can see him now." I told her I didn't want to disturb his meal; I would go by his office someday. She said no, and I should go into the restaurant and see him then. I kept saying no and that I would see him in Manila, but she kept saying I should see him then. She was not being demanding, but encouraging. Finally, she said to her children, "Children, take him in to see Mr.—"

Each took me by a hand and led me to his table. When we got there, they told him I wanted to see him. He and his wife were sitting at a small table eating, and the lady's husband was sitting at a small table next to them.

I told him that I did not want to disturb their meal, but I would see him in Manila. He said that we could talk then and that it would be all right. When I said no again, his wife got up and moved her food and place setting to the table of the other man and asked me to sit down.

I sat down and told him I wanted to talk to him about getting Bibles from his organization. Previously, I had been thinking about how many Bibles it would take to give one to each student in our area, and I came up with roughly 80,000. When he asked how many Bibles I wanted, I told him 80,000. I never kept track of how many

Bibles we used, but after I had been getting Bibles from them for a while, they asked us to be their northern Luzon representative and distributor.

Being Good Stewards

> Therefore submit yourselves to every ordinance of man for the Lord's sake. (1 Peter 2:13)

When I began to get Bibles from CGM, I was entering an area of ministry God had given to them, so I did everything they wanted done the way they directed.

I ministered the way they had set up the program and made sure those who got Bibles and materials from me did the same. I kept the records they wanted kept and filed the reports they asked for, even though it meant extra work.

When we are good stewards, we are given more. I believe that is why we were asked to be their distributor. If we want more, we must be good stewards of what we have.

The number of Bibles I quoted didn't seem to faze him. Next, he asked who I was with, and I told him no one and that the Lord had sent me to the Philippines as an independent. He asked if I had any money, and I told him no and that all I had was what the Lord gave me to live on and the money to do the specific areas of ministry He had called me to.

He explained to me that for a person to get a Bible from their organization, they had to complete a Bible lesson. Basically, it was the four spiritual laws that lead a person to salvation. Since I was used to going into schools and passing out tracts, that was what I wanted to do with the Bibles—give them away. When I explained that to him, he said no. I would have to teach the Bible study. That was the way their program operated, and to be part of it, I would have to do the same.

For me to go back to each school and spend one-and-a-half to two hours in each classroom would mean I would have to spend years going to the schools where I'd already been! It meant I couldn't go to any new schools, yet there were so many schools that needed the Gospel. I didn't realize at the time that the Lord was going to call others to this ministry. But this man was firm, and I'm so glad he was!

Even though the medium of instruction in the Philippines is English, I wasn't sure that the children, particularly the kindergarten and first and second grades, could understand me. A few weeks before Christmas that year, as I went into a classroom to distribute tracts, the teacher asked if I would tell the children a Christmas story. I told her that I couldn't because I had a man waiting for me, and I didn't have the time. At that time, an associate pastor of a local church was transporting me to certain remote places in his tricycle, and he was waiting for me.

There's Always Time to Share the Gospel

As I was walking away from the school, I thought, *What am I doing? Certainly, I have time to tell the children a Christmas story!* No matter what we are doing, we always have time to minister the Gospel of Jesus. I went back in, apologized to the teacher, and ministered the Gospel to the children through a Christmas story. When I asked them if they wanted to receive Jesus and be born again, it seemed to me they all responded. The teacher had them stand up, and we prayed together. Through that, the Lord showed me that even the smallest children could understand me.

It was a wonderful ministry. The workbook they gave us to use was excellent. It was very anointed. We not only used the Bibles and teaching materials but also gave them to other ministries that wanted them, as the Lord led.

The Lord's provision for this new ministry started coming in. I began to go to elementary and high schools. If schools were not remote and difficult to reach, I would go in advance and make an appointment with the principal or head teacher for us to visit that school. On the appointed day, as many of us as were needed would go. It was great, and we really enjoyed it.

One day, we were at Sablan, a small village in the mountains. After we finished our ministry, the principal asked if we would like to have refreshments. As we sat together talking, the principal pointed through an open door at the top of a high mountain peak and said, "There is an NPA camp up there. They came down here last Saturday night to burn down our city hall, but we drove them away." The ones who drove them away would be their Home Defense Force. They are government-trained militia that many villages and towns have for defense.

One day as I was walking through the forest to a school, I saw a stack of freshly carved rifle stocks stacked beside the trail. Years before, the government had declared martial law and confiscated all firearms, so this was not a legitimate enterprise. It's amazing how God's grace enabled me to live and minister in that environment with no fear or concern for my personal safety.

The Lord's Protection

About a month later, I decided it was time to go to Kamoag and Papa schools. Since these are remote schools, I thought they should be taught in the local dialect. I asked a pastor I knew to go with me and do the teaching. That morning, I had to get two of the women set up to do their ministry in two other schools. Because of that, I told the pastor I would meet him at nine o'clock where the road to Kamoag School meets the highway. I had been to that school before to distribute tracts. It was an elementary school of about fifty students. I knew how many Bibles and workbooks to take there, but I had not been to Papa School. I thought it would be about the same size as Kamoag School, but just to be sure we had enough Bibles and workbooks, I brought about twenty extra of each.

Before the pastor arrived, a Philippine army truck packed full of soldiers came up the highway and turned onto the road where I was standing. It was an open truck, so I could see them well.

I was impressed with these soldiers. They were mature and appeared to be well-trained troops, dressed in combat gear with weapons, equipment, and ammunition securely fastened in place. They meant business, and you could see it. When we let the Lord Jesus teach, train, equip, and fully prepare us, that is what we look like to the enemy. They looked pretty cool with their brightly colored bandannas tied around their heads. So do we with Jehovah our Banner flying over us.

The other morning, I saw two jet fighters taking off. As I watched them, I thought about how the Lord sends us out fully armed and ready for battle. Our Commander in Chief sends us to His preplanned targets to destroy the enemy and his works and to set the captives free. I enjoy the fight and the warfare the Lord has called us to. It's a good and just war in which we don't hurt anyone—we just bless them.

As the truck went slowly by, I reached up and gave a handful of tracts to an officer sitting on the passenger's side. I thought, *Maybe he won't give them to the soldiers*. So I reached up a handful to them also.

The pastor arrived shortly after that, and we picked up the shoulder bags full of Bibles and workbooks and started to walk to Kamoag School. For a while, we could see the tire tracts of the army truck, and then they disappeared. We didn't know where they had gone.

At Komoag School, they brought all of the classes together into one big room, and my companion taught them. After he had finished, we ate our lunch and began walking to Papa School. We had no idea where it was, but as we walked, we would occasionally see someone who would give us directions. We were no longer on a road, but were walking through the forest. Once, I saw a tire track in the soft earth and wondered if it was the army truck.

Finally, we broke into a clearing. There we saw two buildings and a couple of houses. One building was the school and the other was now occupied by the soldiers I had seen earlier. We spoke to them as we went by and walked on to the school.

Just as we had entered the clearing, I saw a soldier stepping out of the forest, only a few feet from us. It had started to rain, and he had his poncho on, but he had his gun at the ready. You could see the barrel sticking out from under the edge of the poncho. That's the way we should be, always ready with our weapon, the Sword of the Spirit, the Word of God.

I figured the army had gotten there about four hours before we did and that when they arrived, they went out on a sweep and had then posted sentries.

Papa School was no bigger than Komoag School, so after we finished the ministry at the school, we walked over to the other building and gave the extra Bibles and workbooks to the soldiers. Did they ever appreciate them: They thanked us and thanked us. They would probably be there for a while, and in that remote area with no diversions, they would have plenty of time to read and study.

I once had an opportunity to minister a tract to every person in the Philippine military. I didn't have the money, so I asked a certain woman if her Bible study group would provide the money, and she said yes. But the Lord had told me not to ask for anything, and my visa requirements also prohibited asking for anything, so after I thought about it, I turned it down. How I would have loved to minister the Gospel to those precious people.

It was getting late in the afternoon, and we had walked a long way back into that place, so I asked one of the teachers if there was a shorter way out of there. She said yes and that there was a trail, and she would have one of the children show us. She called a little boy who led us to a trail that led straight down the mountainside. It was steep, rocky, muddy from the rain, and very slippery. We went down, down, down, and finally got to the bottom of a valley where there was a small creek. We crossed the creek and started up the other side. It was so heavily forested that it was hard to see anything ahead of us, but as we climbed up, we saw a highway guardrail. There was the highway. As we climbed up onto the highway, I saw a sign that said, "Sablan."

As I stood on the highway and looked to see where we had come from, I was looking at the mountain that the principal had pointed to and said there was an NPA camp there. When I looked at the mountain, I could see traces of the trail we had come down. It led right up to the top. We had been at or near the NPA camp that day.

I believe the Lord sent the army in there ahead of us to drive out the NPA. He had sent them in early enough so they were well out of the area before we got there. We were constantly in areas infested with NPA, but we were never harmed.

As always happens with the Lord, everyone was blessed. The schoolteachers, the students, the soldiers, and, of course, we were blessed. We gave tracts to everyone we met. My prayer is that when the NPA heard that we were there, and they would, they read some of the literature we left behind and were saved.

I once heard about a testimony an NPA rebel gave. He said that one evening as he sat by a campfire, he picked up a piece of scrap paper to light his cigarette. As he was sticking it in the fire, he noticed there was writing on it and began to read it. It was a tract, and he was born-again.

The Prayers of Others

Again, I believe much of the success and protection I have had over the years has been due to the prayers of others. This is clearly brought out in 2 Thessalonians 3:1–3:

> Finally, brethren, pray for us, that the word of the Lord may run swiftly and be glorified, just as it is with you, and that we may be delivered from unreasonable and wicked men; for not all have faith.
> But the Lord is faithful, who will establish you and guard you from the evil one.

The Lord has provided everything. I believe the only thing that can ultimately stop us is us.

That morning when I got up, I felt ill. I felt like a cold was coming on. I also had a sore knee that hurt when I walked, but I had decided when the Lord called me into the ministry, I would not let anything interfere with what I knew He was directing me to do, not that I've always done that but that is what I've tried to do.

I didn't say anything to anyone. I took these things to the Lord and went on; however, when we got back to the house that evening, the pastor started to make some remarks about the way I had acted that day, inferring that I was pretty soft. We have to be careful of our thoughts and comments to and about others. We hardly ever know what's really going on in someone else's life.

We must never let the devil, people, circumstances, or even ourselves interfere with what God has called us to do. I believe that no matter what, we must do as Jesus did in Luke 9:51 and trust the Lord to provide what we need to accomplish our task:

> Now it came to pass, when the time had come for Him to be received up, that He steadfastly set His face to go to Jerusalem.

We must steadfastly set our faces and go, focusing only on what the Lord has given us to do.

When we went to the high school near Sablan, we experienced hostility from some of the teachers.

There were 300 to 400 students who came from the surrounding mountains. The afternoon we ministered there, as we were getting ready to leave, someone told us that one of the teachers was sick that day, and they had dismissed her section, which was about fifty students. The person said that if we wanted to speak to them, we would have to come back another day. I made arrangements with them to come back the next morning at nine o'clock.

When I arrived the next morning, those students were in class, so I had to wait. When that period was over, they showed me which classroom I was to use, and I began to get everything set up. Before I began, one of the teachers came and told me that one of the classes we had ministered to the day before was free that period and wanted to go back through the ministry again. She asked me if they could and I said yes.

How exciting this was! Here were high school students who could have spent that period loafing, talking to their friends, etc.; but they wanted to sit through ministry they had already been through once. The first time they were a captive audience, but this time they were there because they chose to be. The room was packed that morning. What a blessing!

Remember what I said about the hostility we experienced at that school? There may have been some who were hostile, but look at the multitude who eagerly received the message. We must never be moved by other people's attitudes or actions.

At another high school, we also saw how good God's message is. Upon finishing one room, the people in the office were surprised at the students' conduct. They said that it was the noisiest and most disruptive class in the school. It was a senior class, and some of the boys were already associated with the NPA. They had come down out of the mountains to school only because it was the end of the school term, and they were there to get their certificates. The people in the office were amazed because after the ministry started, everything became quiet.

One boy in particular was disruptive and talking loudly, so the minister told him to be quiet. He seemed surprised that she would say that to him and quieted down. As the teaching continued, all the students got quieter and quieter until there was total silence. By the time the session was over, the boy had tears in his eyes. The only thing that produces results like that is the Word of God, anointed by the Holy Spirit.

I had a situation that was a blessing. When I was in a high school classroom preparing to start the ministry, one of the boys was standing in the middle of the class and when I told him so sit down, he wouldn't. He was probably the class clown, and was looking for attention from his classmates. One thing that I learned about being in the ministry; the enemy will try to resist you every way he can; or every way you will let him get away with. Because of that you have to deal with every situation that comes up, no matter who or where it comes from. If you don't, the devil will run your life, family, and ministry. When I started walking towards him the classroom went totally silent, and everyone's eyes, including his, were as big as a silver dollar. When I got to him, I put my arm around his shoulder, smiled at him, and asked if he didn't think that it would be a good idea to sit down; to which he readily agreed. He could not sit down fast enough. He just needed a little encouragement. The reason I can be merciful, is that there are times that I need a little encouragement too.

Even though we were doing well with the Bible distribution ministry, I returned to distributing tracts because I wanted to finish an area I had started.

CHAPTER 9

God's Unlimited Resources

It was amazing what the Lord was able to do in such a short period of time with so little money. I received about $450 from the United States each month, and the gifts and offerings I received in the Philippines ran about the same.

Because Mrs. Go Belmonte had offered to provide the tracts for one centavo each, meant I could buy 2,600 tracts for one dollar. I was able to give millions of tracts free to other ministries and always had plenty for us. (When I got to 10 million tracts, I quit counting.) The tens of thousands of Bibles we gave over the years, whether through my ministry or to other ministries, for their use or distribution, were always given free. I don't ever remember not being able to give Bibles or tracts to anyone who asked for them, no matter how many they wanted.

Then, there were the millions who we reached through newspapers and television for very little or no money. The Lord has a plan and a way to bring it to pass. As He fed the multitudes with very little when He walked the earth, He can do the same today. He hasn't changed.

> Jesus Christ is the same yesterday, today, and forever. (Hebrews 13:8)

I am not saying that ministries should not have big budgets. Some of the things God has called people to do cost a great deal of money. We should give and sow into good soil according to God's Word and His leading.

Do you remember the rich young ruler in Matthew 19:21 and what Jesus said to him? "Jesus said to him, 'If you want to be perfect, go, sell what you have and give to the poor, and you will have treasure in heaven; and come, follow Me.'"

Notice that Jesus was not telling this man to give up anything; He was telling him to transfer his wealth into heaven and follow Him. Jesus said, "And you will have treasure in heaven." It would still be his; in fact, it still is his. When we give, it is for our benefit as much as anyone else's.

Nothing Restrains the Lord

We might be surprised at how little it takes for us to do what the Lord has called us to do. Remember what the Lord did through Gideon and the 300 men in Judges chapter 7 or what Jonathan said in 1 Samuel 14:6:

> Then Jonathan said to the young man who bore his armor, "Come, let us go over to the garrison of these uncircumcised; it may be that the Lord will work for us. For nothing restrains the Lord from saving by many or by few."

God lives and works in the supernatural. He is not limited to the laws of this natural world, and in Him, neither are we!

I'm not saying we should do or not do anything. Instead, I'm saying we should always look to the Lord and believe that He will do what He has promised.

One day at noon, I went into the office of a school district to get information on the number and locations of the schools in that district. There was no one in the office, so I began looking at some district information that was posted on a wall. As I stood there reading, a teacher came in, and I asked her if she could give me the information I needed. She said that she didn't think that she could, but she said she thought he could and pointed to an elderly man who was sitting at a desk. I had my back turned and had not seen him come in. When she said that, he asked me to come over and sit down.

As I walked toward him, I noticed that his nameplate said he was the district supervisor. Since it was lunchtime, I told him I didn't want to disturb his lunch but would come back later. He said no, it was all right, and we could talk then.

One thing I love about working for and with the Lord is His graciousness. You never have to be pushy, unkind, or inconsiderate to do His work

> Let all things be done decently and in order.
> (1 Corinthians 14:40)

As we talked, I told the supervisor I was going to schools distributing tracts of the Gospel of Jesus Christ and asked if he would give me the names and locations of the schools in his district. As we continued to talk, I told him I would come back the next year, and we would go into the classrooms, minister the Gospel, and distribute Bibles.

When I told him that, he said, "Let's do it now." I told him that I couldn't. I told him that I already had set my schedule for the rest of the school year. But he politely replied, "I want to do it now."

He was a nice, gentle man, and I really liked him, but he was determined to have the Gospel ministered and Bibles distributed in his schools that school year. For about five minutes, I kept telling him we couldn't, and he kept insisting we must do it. Finally, I gave in and asked when he wanted to start. He looked at his calendar and said, "Let's start on February 28," which was about a month away.

As I walked out of his office, I wondered what I had done. I hadn't prayed about it. Was this God's plan? Then, I thought, *I don't know how many students there are. I don't know how many Bibles we need.* I could have gone back into his office and asked him, but I was uncomfortable with that thought. I tried to remember what I had seen posted on the wall. Had I seen anything about total number of students? As I tried to remember, it seemed to me that I had seen something about 3,900 students; so I decided we needed 4,000 Bibles.

Of course, as always, this is something the Lord would have to provide. I had no money for these Bibles.

I was having a difficult time praying for these Bibles since I wasn't sure our doing this ministry was of the Lord. For a couple of weeks, I just prayed about it, but when we started getting close to the 28th, I really got serious.

On the day we talked, the supervisor told me that he would go to every school with us. This did not happen for some reason, but it was all right because the Lord opened the doors for us.

On the second weekend before the ministry was to start, I really prayed. I told the Lord I needed to know something. I told Him the man was getting things ready, and I didn't want to show up the day he was expecting us to start ministry and tell him we were not going to do it.

If we weren't going to do it, I wanted to tell him now before he got everything ready. Again, I asked the Lord to forgive me for not seeking His will before I made the commitment.

I felt bad that I had made this commitment without prayer and had promised the man something before I knew I could do it. As I was waking up Monday morning, the Lord said, "Go and tell CGM what you need." This was legal, since CGM is an American ministry, and telling them would not violate the rules and regulations that governed my visa.

Sharing the Need

When I got there, I went to the office of the woman in charge of that branch of ministry. When I arrived at her office, there was a lady there visiting, and they visited and they visited and they visited. I thank God for His patience with me and in me. Actually, I believe He gave me that waiting time to help settle me into what I had to do.

When the visitor left, I went into her office. I didn't tell her the Lord had sent me. I believed He should be the one who dealt with her. Remember Robin?

I explained I needed 4,000 Bibles for the school ministry, but I didn't have the money. Prior to and after that, I always paid cash when I ordered. Mrs. Go Belmonte once offered me 30-day terms for payment for the tracts. I thanked her and told her it wouldn't be necessary and that I would pay when I ordered.

I always paid as I went unless the Lord directed otherwise, and on those occasions, the money was always there when payment was due.

She responded, "We can't do that. We need the money you give us to help support the ministry." I told her that I understood, thanked her, and began getting up to leave. I did understand; however, I realize now my personal attitude should have been different since I had a Word from the Lord. Not that I should have acted any differently toward her, but I should have had more confidence in what would result as I obeyed His directions.

As I said, I did understand her situation. Twelve thousand pesos is a lot of money. As I was getting up to leave, she stopped me and said, "Tell me again what you are going to do with the Bibles." I explained the ministry to the schools, but she said, "I'm sorry. We have to have the money."

Again, I began to leave, and she stopped me and asked where we were going to do this ministry. I showed her on a map that was hanging on her wall and she said, "That's Ilocano country. I would like to do some ministry there." However, she continued to say that they would need the money.

Every time I tried to get up to leave, she would stop me and ask more questions about the ministry. After a while, she asked, "Can you give 2 pesos per Bible?" I told her no. After that she asked if I could give them 1 peso per Bible, and again, I said no. All I had was enough money to pay the transportation for the Bibles, workbooks, and myself.

She asked if I could give them the money later. I told her all I ever had was what the Lord gave me, and unless He gave me something to give to them, I couldn't give them anything. Since I had no Word from Him about money for the Bibles, I couldn't promise anything. However, I assured her that whatever He gave me for the Bibles, I would be sure to give them.

This exchange went on for about half an hour. After my first statement about my need, I did not say anything more to her about it. I spent the rest of my time answering her questions and trying to leave. All of a sudden she said, "We will give you the Bibles." Praise God!

She told me to go to the warehouse and tell them what I needed. I thanked her, and as I was leaving, I remembered an elementary school principal who had asked if I would come and minister to her sixth-grade class, because they would not be there next year. I had asked how many were in the class, and she told me the number. I remembered it was between forty and fifty, so I asked her for fifty Tagalog Bibles, which she also gave me.

What a wonderful ministry it was! There were nine elementary schools and five high schools in the district. Because five of us were available to minister, we were able to complete all the schools in only about twice the amount of time it would have taken me to distribute tracts.

God's Plan

We found we needed only about 2,000 Bibles, so we took the rest to Baguio. I soon found that the Lord had plans for them too.

One morning, the Lord told me to take some of the Bibles and go to a school in a mining town near Baguio. The whole town was a security zone because of the mine.

When the jeepney got there, the guard at the gate stopped us and spent some time looking closely at each person. When he got to me, he told me that I would have to get out and wait. As I got out, the people helped me unload the Bibles. The guard let the jeepney and the rest of the people go on into town.

Then, he got on the telephone and called somebody. Whoever it was, they talked for quite a while in another language. After he hung up, he told me I could not go in.

He said that I would have to go to some other place or town and get a permit and come back another day.

When he said that, I got very upset, but not with him. He was a polite, accommodating young man who was only doing his job. I was not upset with the man he had talked to. He was also doing his job as he saw it. I was upset with myself for missing the Lord and going out there and wasting all that time.

I thought to myself, *I'm going home and watch television!* I believe emotional fatigue, burnout, was setting in. I stood there fuming and trying to act polite as I waited for a jeepney to come by and take me back to Baguio.

After I stood there for some time, the guard told me he would try again. He got back on the telephone, and after he had talked to someone for a while, he handed the telephone to me and said, "He wants to talk to you." The man on the other end explained that they were expecting a strike at the mine that day, and to enter that area, I would have to get a permit. That meant I would miss this day of ministry.

The Lord Opens the Door

All of a sudden, the man asked, "How long will it take you to do this?" I told him one day. He said, "All right, I will give you one day." Then he talked to the guard, and he admitted me.

When I got to the school, there was a strong anointing. It was amazing how accommodating and helpful those teachers were and how smoothly the ministry went. It was a specially blessed day. One of the teachers was a pastor's wife and already born-again, and she was very helpful to me. I now wonder if she and her husband's prayers were not the difference that day.

During the last class in the morning, I was not able to finish by the lunch break, but the teachers held the class over and told me to take my time and finish. That in itself was unusual. Most teachers and students are anxious to go to lunch on time!

There were two schools at the top of the mountain that I wanted to make arrangements with for ministry to their students. One of them was the high school where the lady ministered to the boisterous class. I decided to go there during the lunch break, that way I wouldn't miss any ministry time.

When I left and came back, I had to go back through security again, and they didn't seem to be very pleased, which kind of surprised me, because of how considerate they had been of my request that morning. I guess to them if I was in for the day, that meant I was in for the day. Looking back, I may have been abusing their kindness to me, which I certainly didn't mean to do.

That afternoon, it was obvious I would not get finished. When school was dismissed, I was still teaching the next-to-the-last class. Of course, they held that class over.

When I finished that class, I began packing to leave. The rest of the school had been dismissed about thirty minutes earlier. I had seen them leave. Then the pastor's wife asked, "What about the other class?" She told me they were holding those students over too.

When I got to that room, the students were waiting quietly for me. It was either the fifth or sixth grade. I thought to myself, *They probably want to be out of here on their way home or out to play.* However, they were all attentive and studious. They were very good.

Sweet Confirmation

The Lord had secured permission for me to be there one day; and He supplied the mercy, grace, and strength we all needed to get everything completed.

I had enough workbooks, but I had not brought enough Bibles, so I made arrangements with the pastor's wife to give Bibles to those who had not received them.

When I returned with the Bibles, she asked, "Do you know what the other teachers said?" Because of the spiritual battles we were constantly fighting, I assumed it was negative. So I asked her pessimistically, "What did they say?"

She told me that they said this ministry was very good. They said, "Others have come here before and passed out Bibles, but they never explained anything. We are going to take what we have learned and tell the Friday afternoon meeting." The Friday afternoon meeting they referred to was a religious meeting for women.

That confirmed to me that we were on the right track. The other thing that blessed me was that these mature, trained, and experienced teachers, who were respected in their community, were now taking the Lord Jesus Christ, whom they had received, to others whom they knew to be in need. The 2,000 extra Bibles were ministered to schools in that area, which is, in keeping with the CGM lady's desire, also an Ilocano area.

Divine Communications

The woman, Mrs. Early, who had bought Bibles for the Bible school in Baguio, had returned to the United States. I had met her husband in Manila, and from talking to him, I knew he was also interested in ministering the Gospel in Asia.

Since she was apparently interested in giving away Bibles, I wrote to them and explained the program I was involved with. I also sent them copies of the workbooks we used and the certificates we issued when students completed the workbooks. I didn't ask them for anything; it just seemed good to let them know about the program.

That same afternoon, I thought about someone else who was interested in ministering Bibles. She was a missionary in her seventies who was in the Philippines at that time. I had met her years before in the United States. She once told me that she was interested in giving Bibles, so I wrote her basically the same letter and explained the program to her. I thought if she were interested, I could send teaching materials and Bibles to her for her ministry.

At the same time, I asked the Lord to give her 1,500 Bibles. About a year later, I saw her at my home church when she was getting ready to return to the Philippines. One thing she said was that she had received enough money to buy 1,500 Bibles. God answers prayer!

Something else happened the afternoon I wrote those two letters. As I said earlier, I believe I was starting to "burnout." Things were bothering me that shouldn't have.

That afternoon, I had gone downtown to make some photocopies. As I was walking back up the hill to the house, the Lord spoke to me strongly and affectionately, saying, "Your God loves you." No one can say, "I love you," like the Lord can! He certainly supplies all of our needs.

The following is an example of God having His own plans and His ability to carry them out through us, if we'll just trust and patiently wait for Him.

> Imitate those who through faith and patience inherit the promises. (Hebrews 6:12)

Because I mailed all my letters to the United States from the hotel, I took the Earlys' letter to Manila on my next trip. My plan was to mail it when I went to the hotel for the Bible study.

I spent that weekend with the family who had the lumber lease. When I got to their home, 100 pesos was all the money that I had. The Bible study was in the afternoon, and that particular Saturday, I had been invited to be part of a wedding that was being held at another hotel that evening.

The man who was getting married was the pastor of a church in northern Luzon. I had known him since I first arrived in the Philippines, and we had been friends since that time.

About a year before the wedding, I had spent a few days ministering in his area. While I was there, I had stayed in the home of one of the families who attended his church. Stan, the husband, worked for the Philippine Department of Forestry and his wife worked for the Department of Agriculture. I got to know them very well and had an enjoyable time staying with them and their children.

I had to take three rides that afternoon and evening. By taxi, they would cost about 50 pesos each. I had to go to the hotel for the Bible study, from there to the other hotel for the wedding and from there back to where I was staying.

I took the letters I wanted to mail to the hotel. It would cost well over 100 pesos to mail them all, but I knew it was possible that the Lord would provide what I needed through a gift or an offering there.

I took a taxi to the Bible study. After the Bible study, we had refreshments, scones and tea and a nice long visit, but no gifts or offerings were given to me, so I couldn't mail my letters.

As I was preparing to leave, the hotel manager told me they would provide my transportation and he called for one of the hotel cars. That meant I had enough money left to get home after the wedding.

When the wedding was over, Stan, the man in whose home I had stayed on my visit to the groom's ministry, walked up and greeted me. As we talked, he said that he was getting ready to leave for the United States to do graduate work. As we continued to talk, I found out he was going to Colorado State University in Fort Collins, Colorado. That's where the Earlys lived, and Mr. Early did some teaching at that University.

The Earlys had lived in the Philippines for many years, and both had done graduate work at the University of the Philippines. They love the Filipino people and know the culture well. They would be perfect people for Stan to meet and become acquainted with, when he arrived. I knew they would do whatever they could to help him.

I told Stan I knew some people in Fort Collins and to wait a minute and I would give him the address. As I went to get the Earlys' letter, to get the address, I got a check in my spirit. Then, I realized that because he is such a quiet and shy man, if I gave him the address, he would never go to visit them. But because he is so accommodating, he would deliver the letter if I asked him to. I took the letter back to where he was and asked if he would deliver it for me, which of course he did. When I visited the Earlys in Fort Collins about a year later, I learned they owned an apartment building and that Stan was renting an apartment from them. It was a blessing to see him at church and to know how well he was doing.

More Miraculous Provisions

During this visit home, I was given a little over $1500 for the Bible ministry and transferred it to the Philippines. Every month, I would call Naomi, Philip, and Maribel, who were managing the ministry, and we would discuss the business affairs and financial needs of the ministry. Month after month, I asked how the Bible money was holding out, and they always told me we had plenty of Bible money left.

I knew we were distributing large numbers of Bibles to other ministries and could not understand why our Bible money had not been used up. Naomi told me they were hauling Bibles into our place by the truck load and out of our place by the jeepney load. A jeepney load would be the equivalent of a pickup truck load.

One day, I finally asked Naomi how we could still have Bible money left, since we were supplying so many Bibles. That's when I was told that CGM had been giving us Bibles for free. Who knows how many Bibles we actually gave away?

Another outstanding answer to prayer happened while I lived in Dagupan. Larry, the medical student from the United States who lived next door, asked me to go to San Fernando to a combination Bible school and orphanage with him. He wanted to treat the children and pass out vitamins and invited me to go along and pray for the children as he treated them.

When we arrived, we learned that the man who had started the Bible school and orphanage had died. He was an American who had been in the Philippines during World War II, and after the war was over, he had returned and begun this ministry. When he died, about half their financial support stopped, and by the time we got there, they were in critical condition.

It was a three-year Bible school with about a hundred students, and there were about fifty children in the orphanage. The staff was mostly Filipino. They were doing all they could to hold things together, but without financial help, they would not make it. I was very concerned about them, but couldn't give them any significant financial help. Not only were they going to go hungry, which would cause a lot of physical problems for them, but also the whole ministry could be lost unless they received substantial, ongoing assistance.

Since I was not able to do much financially, I wrote letters to four ministries in the United States. I described the situation and what the need was, but no one I contacted gave any help. As we continued to pray, the Lord mobilized help and provided their financial and material needs, and the ministry was saved. Through that situation, I was able to clearly see all anyone really needs is the Lord.

A Major Step of Faith

Another ministry also came to know that. After I first arrived in Dagupan City, a local pastor who oversaw a number of churches invited me to minister in one of them. That Sunday morning, I taught that God will supply all our needs. I remember that what I taught was very anointed. After the church service was over, I was invited to go across the road to a church member's home for lunch.

After we were finished eating, the pastor began to tell me about the financial needs of the ministry. He explained that they had a Bible school at that church, and most of the students came from so far away that they could not go home for lunch. He said that most of them were so poor that they couldn't bring a lunch, and his ministry did not have enough money to provide a meal for them.

He said that he and his wife could not help personally, because they had no money. He said that they were living in a small hut with a dirt floor. Then, he asked me for money to help them.

I explained to him that I lived by faith and had no support from anyone except what the Lord provided. I told him the Lord would have to give to me what He wanted me to give to them.

Sometime later, he told me what he and his wife did after I left that afternoon. He said that after they got home, they decided to act on God's word! They prayed together and then went to Dagupan and signed a 1,500-peso per month lease on a place to live and to hold their Bible school.

For a minister and his wife who didn't have 20 to 30 pesos a day to provide meals for their students, this was an outstanding step of faith! The fact that there was such a strong anointing on the message that morning indicated that the Lord was encouraging them to move as He led.

When they took that step of faith, the Lord met them; and the last I heard, four years later, they were still there. The reason the Lord did not want me to supply their needs was because He wanted to! If I were their source, they would look to me, but if He were, they would look to Him.

I'm absolutely convinced there's never a legitimate need that the Lord doesn't have a solution for or will not provide for.

This pastor and his son were the ones who provided Remy and her children's transportation, in his jeepney, when they moved to Baguio. The Lord was continuing to bless him and his ministry. Jeepneys are quite expensive.

I know that at some point, the Lord is going to make a great evangelistic move in Asia. When He does, there is going to be a great need for Bibles, tracts, and Christian literature of all kinds to be printed in the various languages there. For years, the desire had been in my heart to print these Christian materials as they are needed.

One day, Susan, in whose home the tract distribution center was located, told me that the director of an international ministry that had been getting tracts from us would like to meet with me. I agreed, so she set an appointment for us. When we met, he explained that they had been getting tracts from us and wanted to know if we had any other literature available. I told him no; the tract was the only thing we printed and distributed.

The day that I went to that ministry, I had a chance to visit with a young man that was there. He told me that he had been smuggling Bibles into another Asian country. He said that if they were caught, they would spend many years in prison, or could even be killed. He said that the last time that he went, as he was going through customs, he almost got caught. He said that as you enter the customs area there is a long table on each side of the room. Everyone was to lay their luggage on the table, and if you were told to, you had to open for inspection. He said that when he laid his suitcase on the table, the customs lady across the table from him motioned for him to open his bag. He knew that he had, had it. He said that as he started to open his suitcase, a woman in the other line started screaming. When that happened, the customs lady motioned and told him to move on quickly. Some people give up a lot for the Lord.

As we were driving back to the home where I was staying, Susan asked if I would mind stopping to look at a house she was thinking about renting. I told her no. When we pulled up in front of the house, I saw a man standing there. As I stepped out of the cab, before we had even been introduced, he said to me, "Do you know what I have?" Somewhat surprised, I replied, "No, what do you have?" He said, "I have two brand-new printing presses that are the size of a two-story house and can print 10,000 32-page books a day each." He said that they were located in a brand-new, fully equipped printing plant. In addition, he had built a house with a swimming pool on the site. Then, he told me that he had not been able to use this new printing facility for what he had intended, and it had been dormant for three years.

As we continued to talk, he let me know that he would sell the presses to me. Frankly, I believe I could have bought the whole facility. I believe the Lord was giving me the opportunity to print much of the Christian literature that was and will be needed. I'm very sorry to say I didn't take advantage of that opportunity.

Every Need Is Met

After almost three years of my second trip, it seemed to me that the Lord was saying it was time to take a trip home to the United States. I prayed and asked Him when I was to return to the Philippines, and He said September, so I began making preparations.

I would to be gone for three months, so I had to get everything prepared for that period of time. I had to stock Bibles, teaching materials, and tracts. I had to set aside money for rent, utilities, household, and ministry expenses. I also had to set money aside for Naomi's salary and food, since she would be staying at my home taking care of the ministry.

In addition, the law firm was preparing to give me a final statement for the work they had done. And I would need my airline ticket back to the United States, since my return ticket had expired.

As I prepared, the money I needed started to come in. Through the period of about a month, more than 80,000 pesos came in. I never had a month like that before. Except for the $3,000 my home church sent me before I returned to the United States the first time and the 50,000 pesos I received when I returned from the United States, I received about three times as much in the Philippines that month as I ever had before.

CONCLUSION

In conclusion, if there is one incident that sums up the ministry for me, it would be an event that happened one evening shortly after I had moved to Baguio.

Soon after I arrived, I began to go out in the streets and distribute tracts. One Saturday afternoon, I decided to go to the market. It was a cold, rainy day. I spent all afternoon walking through the market giving tracts to shoppers, vendors, shopkeepers, and their employees.

As it was getting dark, I was standing on a dirt street that led out of the market handing out tracts. I was cold and tired, my feet were wet, and I was hungry. I did not know anyone in Baguio very well, so I knew that it would be a lonely weekend.

All of a sudden, everything I was thinking and feeling changed. An elderly lady came past me, and as I gave her a tract, I began to rejoice. She was so old, I wondered if she would ever be back to town again. If she was unsaved, would she ever have the opportunity to have the Gospel ministered to her again? Was this her last or only opportunity?

I stood there rejoicing that no matter what it cost me personally, it was worth everything to have the opportunity to share the Gospel with that lady and all those other people.

Even though I was standing on a cold, lonely mountaintop halfway around the world from home, I was so overcome with joy and gratitude to the Lord for the opportunity He had given me to minister to these precious people that I began to weep. I was touched with His goodness and kindness to me.

> Those who sow in tears, Shall reap in joy.
> He who continually goes forth weeping, Bearing seed for sowing, Shall doubtless come again with rejoicing, Bringing his sheaves with him. (Psalm 126:5–6)

When the Lord called me into the ministry, He said, "Tom, all of these buildings you are building will be burned up some day. If you want the fruit of your labor to last forever, I'm calling you into the ministry. Go into all the world and preach the Gospel to every creature." I'm so thankful to Him for all of the sheaves we are bringing in.

The Lord has obviously blessed me greatly. As it says in Ephesians 3:20, "Now to Him who is able to do exceedingly abundantly above all that we ask or think, according to the power that works in us."

However, if the Lord had not called me into the ministry, I never would have done it. Our Lord Jesus said in John 15:16: "You did not choose Me, but I chose you and appointed you that you should go and bear fruit, and that your fruit should remain, that whatever you ask the Father in My name He may give you."

No one should ever go into the ministry unless God calls them.

I had no personal interest in the ministry. I would have continued to work in business all of my life. If He had not brought me to Tulsa, I would not have come. If He had not told me to start a children's hotline, I would not have done it. I would not have gone to Bible school had He not told me to. I wouldn't have gone to the Philippines, started a school (I couldn't even teach a home Bible study.), written a tract, gone into the schools to minister, or written this book. All this has been His idea. On the other hand, I could never have found or made a life for myself even close to what He has given me.

> God shows personal favoritism to no man. (Galatians 2:6)

Whatever life God has given you and whatever He has called and appointed you to do, He will provide all the love, mercy, grace, miracles, signs and wonders, healings, ability, strength, wisdom, money, favor with man, and everything else to accomplish it. God will withhold nothing you need. He will provide you with as much to accomplish His will for your life as He did our Lord Jesus, Peter, Paul, the other apostles, or anyone else!

I don't believe anyone will ever stand before God and hear Him say, "I'm sorry you were unable to fulfill My plan for your life, but I just couldn't get to you what you needed."

So rejoice! He loves you as much as He loves Jesus (John 17:23). And there is nothing too hard for Him or that He is not willing to do for you.

JESUS SAID, "YOU MUST BE BORN AGAIN"

Now there was a man of the Pharisees named Nicodemus, a member of the Jewish ruling council. He came to Jesus at night and said, "Rabbi, we know you are a teacher who has come from God. For no one could perform the miraculous signs you are doing if God were not with him."

In reply Jesus declared, "I tell you the truth, no one can see the kingdom of God unless he is born again:"

"How can a man be born when he is old?" Nicodemus asked, "Surely he cannot enter a second time into his mother's womb to be born!"

Jesus answered, "I tell you the truth, no one can enter the kingdom of God unless he is born of water and the Spirit. Flesh gives birth to flesh, but the Spirit gives birth to spirit. You should not be surprised at my saying, 'You must be born again.' The wind blows wherever it pleases. You hear its sound, but you cannot tell where it comes from or where it is going. So it is with everyone born of the Spirit:" (John 3:1–8).

Jesus Is Lord, Jesus Is God

A week later His disciples were in the house again, and Thomas was with them. Though the doors were locked, Jesus came and stood among them and said, "Peace be with you!" Then He said to Thomas, "Put your finger here; see my hands. Reach out your hand and put it into my side. Stop doubting and believe:"

Thomas said to him, "My Lord and my God!" Then Jesus told him, "Because you have seen me, you have believed; blessed are those who have not seen and yet have believed." (John 20:26–29)

An angel of the Lord appeared to him in a dream and said, "Joseph son of David, do not be afraid to take Mary home as your wife, because what is conceived in her is from the Holy Spirit:" [Jesus was conceived in Mary, not by a man but by the Holy Spirit of God.] "She will give birth to a son, and you are to give him the name Jesus, because he will save his people from their sin." All this took place to fulfill what the Lord had said through the prophet. "The virgin will be with child and will give birth to a son, and they will call him Immanuel"—which means, "God with us." (Matthew 1:20b–23)

Jesus Is God with Us

But about the Son he says, "Your throne, Oh God, will last forever and ever, and righteousness will be the scepter of your kingdom. You have loved righteousness and hated wickedness; therefore God, your God, has set you above your companions by anointing you with the oil of Joy." (Hebrews 1:8–9)

Twice in this scripture, God the Father calls Jesus Christ God.

The reason that Jesus Christ came to this earth and died on the cross was to pay the penalty for our sins. "For the wages of sin is death, but the gift of God is eternal life in Christ Jesus our Lord" (Romans 6:23). Jesus was paid the wages for our sins so that we may have eternal life.

> But now a righteousness from God, apart from law, has been made known, to which the Law and the Prophets testify. This righteousness from God comes through faith in Jesus Christ to all who believe. There is no difference, for all have sinned and fall short of the glory of God, and are justified freely by his grace through the redemption that came by Christ Jesus. God presented him as a sacrifice of atonement, through faith in his blood. (Romans 3:21–25a)

Every person has sinned, and the penalty for that sin is hell, which is spiritual death, eternal separation from God.

> For God so loved the world that he gave his only begotten Son that whoever believes in him shall not perish but have eternal life. For God did not send his Son into the world to condemn the world but to save the world through him. (John 3:16–17)

God knows that each person is going to hell forever because each person has sinned. But God loves you so much that he sent Jesus to pay the penalty for your sins, so that you don't have to. God wants you to have eternal life in heaven with him. Jesus did not come to earth to condemn you for your sins. He came to save you from eternal death in hell. Jesus said in John 12:47b, "For I did not come to judge the world, but to save it."

How can you receive this eternal salvation through Jesus Christ? Jesus said, "Here I am! I stand at the door and knock. If anyone hears my voice and opens the door, I will come in and eat with him, and he with me" (Revelation 3:20). Jesus is standing and knocking at the door of your heart right now. Will you open your heart and let him come in and forgive you of your sins and give you eternal life in heaven?

"For it is by grace you have been saved through faith and this not from yourselves, it is the gift of God, not by works, so that no one can boast" (Ephesians 2:8–9). Salvation through Jesus is a free gift from God our Father; you don't have to do anything to earn it.

Jesus also said, "I am the way and the truth and the life. No one comes to the Father except through me" (John 14:6). There is no other way to God our Father and eternal life in heaven except through Jesus Christ.

> If you confess (speak out freely) with your mouth, "Jesus is Lord"; and believe in your heart that God raised him from the dead, you will be saved. For it is with your heart that you believe and are justified, and it is with your mouth that you confess and are saved. As the Scripture says, "Anyone who trust in him will never be put to shame." For there is no difference between Jew and Gentile—the same Lord is Lord of all and richly blesses all who call on him, for "Everyone who calls on the name of the Lord will be saved." (Romans 10:9–13)

"And this is the testimony: God has given us eternal life, and this life is in his Son. He who has the Son has life; he who does not have the Son of God does not have life" (1 John 5:11–12).

So if you want to become a child of God and be born again into the kingdom of God, pray the following prayer and God will forgive your every sin. There is no sin so great that the blood of Jesus cannot wash you completely clean.

> My Father in heaven, I come and ask You to forgive all of my sins now so that I can be born again and spend eternity in heaven with You. I believe in my heart that You raised Jesus from the dead, and I now confess with my mouth that "Jesus is Lord." Thank you Father, for my salvation. I believe that I'm born again and saved and a child of God. In the Name of Jesus Christ my Lord, I pray. Amen

If you prayed this prayer, you are now a child of God with eternal life. "Yet to all who received him, to those who believed in his name, he gave the right to become children of God - children born not of natural descent nor of human decision or a husband's will, but born of God" (John 1:12–13).

This tract may be reproduced for distribution.

Scripture taken from the Holy Bible, New international Version.® NIV® Copyright© 1973, 1978, 1984 by International Bible Society. Used by permission. All rights reserved.

ABOUT THE AUTHOR

The author always makes it his aim to be well pleasing to God. The Heavenly Father said of the Lord Jesus Christ, "This is My beloved Son, in whom I am well pleased" (Matthew 3:17). Before Jesus began His ministry, He was beloved and well pleasing to God.

Before David was crowned king of Israel and began the work that God had given him to do, He said, "I have found David the son of Jesse, a man after My own heart, who will do all of My will" (Acts 13:22b). While David was still a shepherd boy, God said of his heart, "For the Lord does not see as man sees; for man looks at the outward appearance, but the Lord looks at the heart" (1 Samuel 16:7b).

The Bible says of Moses before he began his ministry, "By faith Moses, when he became of age, refused to be called the son of pharaoh's daughter, choosing rather to suffer affliction with the people of God than to enjoy the passing pleasures of sin, esteeming the reproach of Christ's greater riches than the treasures of Egypt; for he looked to the reward" (Hebrews 11:24–26).

Then, there was Caleb, who was willing to go wherever God sent him and do whatever He wanted him to do. God said of him, "But My servant Caleb, because he has a different spirit in him and has followed Me fully, I will bring into the land where he went, and his descendants shall inherit it (Numbers 14:24).

So rejoice, whether God calls you into the ministry or not, you can be beloved and well pleasing to Him.

CPSIA information can be obtained
at www.ICGtesting.com
Printed in the USA
FFHW021935050119
50052226-54861FF